Realism, Mathematics and Modality

For Julia and Elizabeth

Realism, Mathematics and Modality

HARTRY FIELD

BLACKWELL
Oxford UK & Cambridge USA

Blackwell Publishers
108 Cowley Road, Oxford, OX4 1JF, UK

3 Cambridge Center
Cambridge, Massachusetts 02142, USA

British Library Cataloguing in Publication Data
A CIP catalogue record for this book is available from the British Library.

Library of Congress Cataloging in Publication Data
Field, Hartry H., 1946–
    Realism, mathematics, and modality / Hartry Field.
        p.   cm.
    Includes bibliographical references and index.
    ISBN 0–631–18087–7
    1. Mathematics—Philosophy.   2. Space and time.   I. Title.
QA8.6.F54   1991
510'.1—dc20
                                                    91–14952
                                                         CIP

Typeset in 10 on 12 pt Garamond
by Photo·graphics, Honiton, Devon
Printed in Great Britain by
T.J. Press (Padstow) Ltd, Padstow, Cornwall.

Table of contents

Preface

This volume collects together all my papers in the philosophy of mathematics, and one paper on the philosophy of space and time that is closely connected with the other papers in several of its themes. Most of the papers have been unaltered except in trivial ways, but I have made some substantial changes in 'Is mathematical knowledge just logical knowledge?' and in the title essay. A brief description of the changes is included at the end of the opening footnote in each of these essays. I have also included postscripts to three of the articles, and a long introductory essay to the volume as a whole. Aside from this introductory essay, the papers are printed in the order in which they were written.

The essays originally appeared in the following places:

'Realism and anti-realism about mathematics' in *Philosophical Topics* 13 (1982): pp. 45–69.

'Is mathematical knowledge just logical knowledge?' in *Philosophical Review* 93 (1984): pp. 509–52.

'On conservativeness and incompleteness' in *Journal of Philosophy* 81 (1985): pp. 239–60.

'Platonism for cheap? Crispin Wright on Frege's context principle' in *Canadian Journal of Philosophy* 14 (1984): pp. 637–62. (It was there entitled 'Critical notice of Crispin Wright: *Frege's conception of numbers as objects*'.)

'Can we dispense with space-time?' in P. Asquith and P. Kitcher, eds, *PSA 1984: Proceedings of the 1984 Biennial Meeting of the Philosophy of Science Association* vol. 2 (1985): pp. 33–90.

'Realism, Mathematics and Modality' in *Philosophical Topics* 19 (1988): pp. 57–107.

I would like to thank the editors of these journals and volumes for permission to reprint. And thanks to *The New Yorker* for the permission to reprint Rea Irvin's cartoon.

I gave a seminar in the philosophy of mathematics at the University of Southern California in the spring of 1988, and some of the additional material for this volume was discussed in it. Several of those in attendance made comments which helped me to improve the material: in particular, Jonathan Barrett, Gawayne Batchelor, Stuart Cornwell, Tomoji Shogenji, and Takashi Yagisawa. Brian Loar read an early draft of part of the Introduction and, as usual, gave me good advice.

A NOTE ON THE PAPERBACK EDITION

Some changes have been made on pages 105, 106 and 110 to correct a small error. Also there are minor stylistic changes on pages 43 and 199.

1

Introduction: fictionalism, epistemology and modality

A number of themes run through the papers collected in this volume, but certainly the most dominant is the idea that we can develop a satisfactory fictionalist account of mathematics. I characterize this fictionalist position in section 1 of this introduction, and discuss some of the considerations which motivate it, and some which make it difficult to achieve, in sections 2–4. Section 7 addresses what has been perhaps the most common worry about the form of fictionalism I have defended. Sections 5 and 6 discuss a second theme prominent in several of the essays in the volume: that modality has some role, but only a very limited role, to play in ontological discussions.

PART ONE

1 Fictionalism

A mathematical realist, or platonist, (as I will use these terms) is a person who (a) believes in the existence of mathematical entities (numbers, functions, sets and so forth), and (b) believes them to be mind-independent and language-independent. So there are two different (at least, verbally different) kinds of anti-platonist position. One involves a disbelief in mathematical entities; the other takes the idealist position that mathematical entities exist but only as some sort of 'mental construction' or 'construction out of our linguistic practices'.[1] It is not

---

[1] An example of the linguistic variant of idealism: 'The existence of numbers is just constituted by the fact that there is a legitimate practice involving discourse with a certain structure, and that certain products of this discourse meet the standards of correctness that it sets' (Stalnaker 1988). Stalnaker describes his position as platonist, but this does not seem like Platonism as it is normally understood.

immediately obvious that the idealist version of anti-platonism ought to be regarded as importantly different from the simple denial of mathematical entities – just as it is not immediately obvious that Berkeleyan idealism should be regarded as importantly different from the denial of the existence of tables and chairs and the like. I will not pursue this issue here. I will make a few very brief remarks about the idealist form of anti-platonism in section 4 of this introduction, but for the most part I intend to ignore it.

If we ignore mathematical idealism, then we can say that anyone who adopts an attitude of literal belief toward mathematical theories taken at face value is a mathematical realist. For a mathematical theory, taken at face value, is a theory that is primarily about some postulated realm of mathematical entities: numbers, or functions, or sets or whatever (or some combination, like numbers and sets together).[2] You can't consistently believe the theory without believing in the entities it postulates.

This suggests that (if we continue to ignore the possibility of mathematical idealism), an anti-platonist should embrace fictionalism about mathematics – or at least, fictionalism about mathematics-taken-at-face-value. A fictionalist about mathematics-taken-at-face-value is someone who does not literally believe mathematical sentences, at least when they are taken at face value. (Or, if you prefer to 'semantically ascend', a fictionalist is someone who does not regard such sentences, taken at face value, as literally true.) The fictionalist *may* believe that there is some non-face-value construal of mathematical sentences under which they come out true; he or she may even believe that some such construal gives 'the real meaning of' the mathematical sentence, despite its departure from what the mathematical sentence appears to mean on the surface. My own view, though, is that the second of these additional claims is an uninteresting verbal one insofar as it goes beyond the first; and that the first, though of some interest, unnecessarily constricts the fictionalist. (I'll have a few brief remarks on the latter point at the end of this section.)

A fictionalist needn't (and shouldn't) deny that there is *some* sense in which '2+2=4' is true; but granting that it is true in some sense does

---

[2]   The theory may mention non-mathematical entities too, for instance if it says that for every entity, even a non-mathematical one, there is a set with that entity as its sole member. But I would not count it as mathematical unless it postulated or purported to refer to some mathematical entities. This is a necessary but not a sufficient condition for being a mathematical theory: a theory of mathematical physics that uses mathematics to describe the physical world is not an example of what I am calling a mathematical theory, for it is not the kind of thing that any mathematician would claim to believe on 'mathematical evidence' alone.

not commit one to finding any interesting translation procedure that takes acceptable mathematical claims into true claims that don't postulate mathematical entities. Rather, the fictionalist can say that the sense in which '2+2=4' is true is pretty much the same as the sense in which 'Oliver Twist lived in London' is true: the latter is true only in the sense that it is true *according to a certain well-known story*, and the former is true only in that it is true *according to standard mathematics*. Similarly, the fictionalist *believes* that 2+2=4 only in the sense that he or she believes that standard mathematics *says that* (or, *has as a consequence that*) 2+2=4;[3] just as most of us believe that Oliver Twist lived in London only in the sense that we believe that the novel says that or has as a consequence that Oliver Twist lived in London. If one believes only this, it seems rather natural to say that one does not literally believe that Oliver Twist lived in London (after all, one doesn't believe that if one had gone to London in the nineteenth century one would have found Oliver there); similarly, the fictionalist who regards the comparison as reasonably apt will find it natural to say that he doesn't literally believe that 2+2=4. (He or she can still advise a young school-child to say that it's true on a true–false test, of course; similarly in the Oliver Twist case.)

I am strongly inclined to think that the fictionalist view of mathematics is correct, though I have to acknowledge that there are *prima facie* obstacles that the fictionalist needs to find a way to overcome. But for now let us defer arguments against fictionalism, and focus only on what the fictionalist view *is*.

One natural question about fictionalism is this: surely the fictionalist must grant that standard mathematics is in many ways a *good* 'story': but how could he or she give any content to this except by saying that its goodness consisted in its being true? Of course, there is a dismissive reply: that the Oliver Twist story is a rather good one too, but that *its* goodness doesn't consist in its truth. But this is hardly satisfying by itself: obviously, the way in which the 'story' told by standard

---

[3] For impure mathematical claims it's slightly more complicated: the fictionalist believes that 2 is the number of planets closer than the Earth to the Sun in the sense that he or she believes that this follows from standard mathematics *together with purely non-mathematical facts*: among the purely non-mathematical facts is the fact that $\exists x \exists y \{x$ and y are planets closer than the Earth to the Sun; $x \neq y$; and for any planet z closer than the Earth to the Sun, either $z=x$ or $z=y\}$. I take 'standard mathematics' to include a 'bridge law' that connects up such 'statements of non-mathematical facts' to their more mathematized counterparts. (I take 'the non-mathematical facts' here to involve no claim about mathematical entities, *not even that such entities don't exist*: otherwise, the possibility of an inconsistency between standard mathematics and the non-mathematical facts would threaten the above account.)

mathematics is good is very different from the way in which the Oliver Twist story is good. There are a number of differences that might be mentioned here, but perhaps the most important is that mathematics is good *as an instrument* that can be applied in domains outside mathematics, and nothing like this is true of the Oliver Twist story. (When one wants to stress this difference, one uses the word 'instrumentalist' instead of 'fictionalist' in the case of mathematics.)[4] Because of this difference, there is certainly room for initial suspicion that for standard mathematics the only reasonable account of goodness involves truth – or, perhaps, necessary truth.

I think, though, that the fictionalist can provide an alternative account: it was developed in Field 1980 and elaborated in several of the essays below. On this account, truth isn't required for goodness (so necessary truth isn't required either); what is required instead is something called conservativeness, which embodies some of the features of necessary truth without involving truth. I don't want to say that conservativeness is the *sole* virtue in a mathematical theory – there are secondary virtues, such as interestingness, elegance, the having of applications outside mathematics and so forth, and these secondary virtues are quite important. (A believer in mathematics recognizes these secondary virtues too, in addition to the primary virtue of truth or necessary truth.) But I contend that none of the virtues requires that the mathematical theory be true. It may be, of course, that conservativeness is too weak a virtue to account for the applicability of mathematics to the world – that we need truth in addition – but now I am only explaining what the fictionalist view is, not saying that it is correct.

I have heard two arguments for the unintelligibility of fictionalism about mathematics that deserve comment. Argument one is that it is unintelligible to deny the truth of mathematical assertions like '2+2=4', since it is simply a consequence of the meaning we have assigned to '2', '4', '+', etc., that this and similar assertions hold. It seems to me, though, that this can't be right. There *is* a *somewhat* plausible claim that underlies it: that our number-theoretic concepts are such that we wouldn't count anything as the number 2 unless there were also something we counted as the number 4 and unless we recognized an addition function mapping the thing we counted as 2 and itself into the

---

[4] Mathematical instrumentalism is sometimes differentiated from mathematical fictionalism: the instrumentalist is said to hold that mathematical claims lack truth value, while the fictionalist is said to hold that they are false. In my opinion this is a totally uninteresting difference; the important point to both positions is that the acceptability of a mathematical claim is in no way dependent on any truth value it may have. In the case of mathematics (and also in the case of physics), where the applicability is evident, I will use the words 'fictionalism' and 'instrumentalism' interchangeably.

thing we counted as 4. In other words, the claim 'If there are numbers then $2+2=4$' has some claims to count as an analytic truth, indeed one so obvious that its denial is unintelligible. In a similar way, perhaps, nothing would count as Santa Claus unless it were human, lived at the North Pole, flew reindeer every Christmas Eve, etc. – or at least, did most of those things. But it can't be analytic (or, a purely conceptual truth) that there *is* a Santa Claus that lives at the North Pole and flies reindeer, and it can't be an analytic or purely conceptual truth that there *are* objects 1, 2, 3, 4 etc. obeying such laws as that $2+2=4$. An investigation of conceptual linkages can reveal conditions that things must satisfy if they are to fall under our concepts; but it can't yield that there are things that satisfy those concepts (as Kant pointed out in his critique of the ontological argument for the existence of God).

Argument two is that it is simply unintelligible to say that one regards standard mathematics as good, and believes that $2+2=4$ is true according to standard mathematics, but yet one doesn't *believe* $2+2=4$ (or standard mathematics); what more could there be to *believing* standard mathematics than *regarding it as good*? The Oliver Twist and Santa Claus comparisons are inept, the argument continues. For in the case of the sentence 'Oliver Twist lived in London', one can give clear content to the idea that one only believes that this holds *in the story*: the fact that one doesn't literally believe the claim itself comes out in such facts as that one wouldn't expect to find Twist listed in the records of the London orphanages of the period. But, the argument goes, no analogous content can be given to merely believing that $2+2=4$ holds *in the story*, rather than strictly believing that $2+2=4$. One can't give content to the claim that one holds a fictionalist view by saying that one denies that an inspection of the universe will turn up any mathematical entities, for of course all but the most mystical platonist would agree that mathematical entities are not available for that kind of inspection.

In reply, it is first worth noting that even if argument two succeeds, it does not undermine what might be called 'weak fictionalism': the doctrine that mathematical claims are true only in the way that fictional claims are true. Rather, argument two is really an argument that any sensible form of platonism *agrees* that mathematical claims are true only in the way that fictional claims are true; what is wrong, according to the argument, is really only 'strong fictionalism', that is, weak fictionalism coupled with the doctrine that weak fictionalism and platonism are to be distinguished.

But this point aside, I do not think that this argument that weak fictionalism and platonism coincide is correct. At least, it does not seem to me correct as applied to the version of fictionalism that I incline towards, the kind which I have defended in many of the papers in this volume. For there are clear substantive differences between fictionalism of that sort and platonism: in particular, the fictionalist differs from the platonist in regarding different theoretical questions as important. More fully, fictionalism of the sort I recommend has both positive content and negative content. Its positive content is that it commits one to abjuring all appeal to mathematical entities in explanations when the chips are down: it must be possible, for instance, to develop theoretical physics without any appeal to mathematical entities. The claim that such an elimination of mathematical entities from explanations is possible is a very substantial claim; there are serious difficulties in defending it, difficulties which have led most philosophers to think that this sort of fictionalism is unworkable. (I think that there are also benefits of showing how to eliminate mathematical entities from explanations, independent of issues about fictionalism – see the end of section 3 of this introduction, and section 5 of essay 6.) Whatever the merits of the eliminability claim, that claim is the positive content of the form of fictionalism I recommend. The negative content of fictionalism is that it avoids having to answer some questions that seem to need answering on a platonist view. This is the main *motivation* for fictionalism, and I will have more to say about it shortly. It seems to me a legitimate matter of dispute whether, at the present stage of knowledge, it is a good exchange to shift the important questions in the way that (this sort of) fictionalism recommends; my present point is only that for better or worse fictionalism does license such a shift, and this shows that contrary to argument two it is an intelligible position genuinely distinct from platonism.[5]

Fictionalism about mathematics is one alternative to platonism. Now, much as phenomenalists have claimed that rejecting matter didn't prevent them from literally believing that they were sitting at a table, some anti-platonists say that rejecting platonism doesn't prevent them from literally believing that $2+2=4$ or that there exists a number between 15 and 17; those claims are literally true, the anti-platonist can hold, as long as they are construed in the appropriate fashion. (And 'the appropriate fashion' of construing them needn't be an idealist one: it could instead

---

[5] There may be some dispute as to whether fictionalism really does differ from platonism in 'negative content': one line of defence of platonism has been to argue that its commitments are less than the anti-platonist contends (though I have never seen this argued very convincingly). But at least by virtue of its positive content, a fictionalism of the sort I defend is substantively different from platonism. Against a form of fictionalism like van Fraassen's (1975, 1980) that would avoid the hard work of showing how mathematical entities are eliminable from explanations, the argument of the previous paragraph has a better chance of success.

be, for instance, a modal construal that involved no reference to any objects other than physical ones, as suggested in Putnam 1967b or Hodes 1984b or Chihara 1984.) In advocating fictionalism, I do not really mean to be opposing such views (though I do mean to refrain from endorsing them): for such views *are* fictionalist about mathematics *taken at face value*, and I have not committed myself to a fictionalism more radical than that.

I should add as an aside, though, that I think that there are difficulties in trying to force anti-platonism into the 'non-standard construal of mathematical theories' mode. One problem arises from the need to give an account of the applications of mathematics to the physical world. In the case of number theory, perhaps, the problems are not so great: for there is one specially central kind of application of number theory to the world, one that works by using numerals to 'encode' numerical quantifiers (see Hodes 1984b); and provided one is willing to utilize a rich enough logic,[6] one can give a general translation procedure for number-theoretic claims that gives them a natural role to play in these applications. But in other cases, such as the theory of differentiable functions of a real variable, there seems to be no single canonical application: one can apply such mathematics to physical space, to degrees of belief, and to much else. I think that this makes much more difficult the problem of finding a translation procedure for the mathematical theory that gives rise to a natural account of its applications. I will not pursue this, though. (I have exhibited an approach to applications elsewhere that does not rely on any translation procedure of the mathematics: it is in Field 1980, but there are some sketchy remarks about it in several of the essays below. Sections 6 and 7 of essay 7 discuss special difficulties for using modal translations of mathematics in applications.) It also seems to me doubtful that forcing one's anti-platonism into the mode of a nonstandard construal of mathematics has benefits that compensate for the difficulties: on this point, see section 8 of essay 7. But there is no need to press this point: the nonstandard

---

[6] Hodes uses impredicative second order logic and a certain sort of not-purely-logical modality. I am dubious about both; in particular, his view that the former is available to the anti-platonist rests on a sharp distinction between 'objects' on the one hand and 'properties' (or 'Fregean concepts', as Hodes prefers) on the other, and on the definition of platonism as the view that there are mathematical *objects*. To me this definition of platonism seems perverse, for the epistemological and other difficulties that mathematical objects like sets have been thought to raise seem to arise with equal force for the mathematical 'non-objects' (properties or Fregean concepts) that are in the range of the second order variables, and there are other problems there as well (such as the mysteries surrounding Fregean 'unsaturatedness'). But perhaps there are less ontologically loaded logics that would have served Hodes' purposes as well.

construal approach differs from a more thorough fictionalism only in a point of strategy; the anti-platonism is what's important.

2 Initial Plausibility

Fictionalism is often portrayed by platonists as a radical position, quite at odds with the views of the average non-philosopher. I rather doubt that this is so. I don't think it at all obvious that the average person who calculates, or the average physicist who quantifies over mathematical entities while theorizing, or even the average mathematician, literally believes that there are mathematical entities. The average non-philosopher, I suspect, has not thought enough about what platonism involves and what fictionalism involves to have anything like a consistent view of the matter.

But whatever the views of the average non-philosopher, there are considerations that appear to favour the platonist, and which the fictionalist will have to deal with. These considerations have to do with the fact that mathematics is not just an autonomous discipline. Rather, mathematics has many applications outside mathematics – in scientific explanation, in the description of our observations, in metalogic, and in many other areas. It is the fact that mathematics appears indispensable in applications (indispensable without incurring high costs, that is) that provides the main source of arguments for platonism.

In several essays below I have gone further: I have said that the *only* serious arguments for platonism depend on the fact that mathematics is applied outside of mathematics. These remarks (which I stand by) have led quite a few people to accuse me either (i) of ignoring the possibility that mathematics is known by investigating conceptual interconnections, or (ii) of ignoring the reasons that mathematicians actually give for their mathematical beliefs. As for (i), I have already noted in the previous section ('argument one') that reflection on conceptual interconnections cannot yield knowledge of the existence of mathematical entities: it can at best yield knowledge that *if* there are entities that we can correctly call mathematical, *then* they obey the usual mathematical laws. (ii) deserves a more extended discussion, but I will now argue that it too is incorrect.

There are two types of arguments that mathematicians tend to give for their mathematical 'beliefs' – I use the quotes since as remarked there is a good deal of question in my own mind as to whether the typical mathematician literally believes the sentences of standard mathematics. First, the mathematician may argue for a claim by proving it (perhaps sketchily). But what is called rigorous mathematical proof is really just logical derivation of the claim being proved from other mathematical claims (and what is called *sketchy* proof is *sketchy* logical derivation from other mathematical claims). The fictionalist and the platonist agree that logical derivations are to be trusted. The epistemological question about mathematics is not about these logical derivations, but about the mathematical claims used as premises of the derivation: unless they are believed (not just accepted as part of an interesting story), the derivation obviously offers no ground for literal belief in the claim derived. The second kind of argument that mathematicians typically offer for accepting a mathematical claim is that the claim has attractive consequences. Again the fictionalist agrees that it has those consequences, and is likely to agree that the fact that it has those consequences is good reason to accept it for many purposes; the fictionalist *might* indeed go further, and agree that if the attractive consequences are to be literally believed, then the argument gives some sort of hypothetico-deductive reason for literally believing the claim in question. But even this (perhaps dubious) concession does not give grounds for believing the claim in question unless those attractive mathematical consequences are themselves literally believed; and a fictionalist mathematician will refrain from literally believing them.

But, it will be said, I am ignoring the fact that mathematicians (and ordinary people) find many mathematical claims *initially plausible*; plausible *independently of argument*. Well, perhaps they do and perhaps they don't. I certainly grant that mathematicians and ordinary people find many mathematical claims *natural to accept* independently of argument, where 'accept' means simply 'incorporate into one's mathematical story'. There are obvious reasons why some mathematical claims are very natural to accept in this sense, and these reasons don't presuppose platonism. The reasons for finding certain mathematical claims natural (even if not literally believable) will vary somewhat from one mathematical claim to another. For instance,

(a) {Human females} ∪ {human non-females} = {humans}

is natural to accept largely because of its intimate association with the logical truth '$\forall x(x$ is human if and only if either x is human and female or x is human and not female).' Similarly,

(b) $1+1=2$

is natural to accept largely because of its intimate association with logical truths like 'If there is exactly one apple on the table and exactly one green thing on the table and no apple on the table is green then there are exactly two things on the table which are either apples or green'; here 'there is exactly one' and 'there are exactly two' are numerical quantifiers, definable in terms of ordinary quantifiers plus identity. (The fact that (b) is intimately associated with certain sentences involving numerical quantifiers is, of course, what lies behind most standard applications of the natural numbers to reasonings about the physical world. The fictionalist view can easily make sense of these applications: see essay 2 and the postscript thereto.) The situation is similar with

(c)   Between any two real numbers there is another real number;

this is natural to accept largely because of its intimate association with an analogous claim about points on a line in physical space. (Again, this intimate association is involved in some though not all of the most familiar applications of the real numbers.)

(d)   For any physical objects x and y, there is a set containing x and y as its only members

draws some of its naturalness from the claim that there is an *aggregate* of x and y; admittedly, the claim about aggregates gives direct naturalness to (d) only in the case when x and y don't overlap, but the extension to the case where they do overlap is also natural because of its simplicity. And in each of cases (a)–(d), part of the naturalness of the claim doubtless comes from our mathematical education: taking (b) as an example, we are taught from early childhood both that $1+1=2$ and that we can pass freely between numerical quantifications and corresponding claims about numbers.

There are of course mathematical claims that unlike (a)–(d) are not intimately connected with non-mathematical claims, and which seem natural even when we first hear them, so that education is not directly a factor: an example is

(e)   There are inaccessible cardinals.

But a fictionalist will find these natural too: they are natural ways to extend the 'story' of Zermelo–Fraenkel set theory. (We find some extensions of ordinary pieces of fiction more natural than others, so why not in mathematics?) It seems to me quite tendentious to take the uncontroversial fact that some mathematical claims are, for diverse reasons, natural to accept, and redescribe it in terms of their being 'initially plausible'.

But suppose that most mathematicians do find these claims initially plausible, rather than just natural. What follows? Well, one can try to make it look like a lot follows by drawing an analogy between initial plausibility judgements in mathematics and perceptual judgements about the physical world. There is some plausibility to the analogy: after all, part of what makes a claim like 'This is red' perceptual is that it is believed *independent of argument* in the appropriate circumstances, so it too could be said to be 'initially plausible' in those circumstances. Given the perceptual analogy, should we not say that mathematical claims like (a)–(e) are in as good epistemological shape as perceptual judgements about ordinary objects?

There are two reasons to doubt this. One, which I shall *not* discuss now, is that there seems to be a crucial difference between the cases: the difference arises from the fact that there are unproblematic connections (typically causal connections) between what we perceive and our perceptual judgements, whereas there are no such unproblematic connections in the case of plausibility judgements in mathematics. (Some brief remarks on the significance of this difference can be found in section 4 of this introduction, and in section 2 of essay 7.) But the other reason for doubting the moral suggested at the end of the previous paragraph is that it rests on a very naive view of the epistemological significance of perception.

Whatever the defects of coherence theories of knowledge, one thing that seems right about them is that claims of 'direct perception' never have unchallengeable status. More specifically, not just individual perceptual judgements but whole practices of making perceptual judgements of a certain sort can be questioned, *when an alternative proposal for a perceptual practice is available*.

Here's an example (deriving from Feyerabend 1975). It may well be that the epistemic practices of astronomically ignorant people licensed the assertion 'The sun is rising' in the observational circumstances typical of the early morning, and that this assertion was always understood literally, as meaning that the sun was in absolute upward motion. But it is hard to deny that when the astronomical theory implicit in this observational practice was undermined, it became rational to modify the practice (either by modifying the words used to report observations; or by understanding these words differently; or by using the same words understood the same way, but no longer literally believing the observational reports and believing instead only that they are a useful though literally false way of conveying the truth about the change in angle between the horizon and the sun). *It seems to be part of our general methodology to consider alterations of our perceptual practices under certain circumstances: in those circumstances, we compare the old perceptual practice with a proposed new one, and make the alteration if it seems to lead to better results.*

A platonist might respond as follows: 'All this analogy suggests for the mathematical case is that we shouldn't hold on to certain types of mathematical plausibility judgements after they have come into conflict with other mathematical plausibility judgements or with perceptual judgements. But mathematical judgements never do come into direct conflict with perceptual judgements, and if we only revise mathematical plausibility judgements when they conflict with other mathematical plausibility judgements we will never be led to a fictionalist position.' But a platonist who so responds has failed to appreciate the force of the example. What happens in cases like the above is *not* that the old observational practice as a whole undermines a particular part of that practice (the part I've cited), by lending support to an astronomical theory that contradicts that part of the observational practice. For the fact is that the new astronomical theory could never be thought of as well-supported as long as one relied on the old perceptual practice. (The old observational practice leads so immediately to an astronomical theory in which the sun is in absolute upward motion in the morning that it is hard to see how a simple amalgamation of initial credibilities from the totality of our observations can lead to a new theory that radically conflicts with this.) What happens in the example is rather that the mere *suggestion* of an alternative perceptual practice together with some sketch of what life would be like if we accepted that alternative practice is enough to cast enough doubt on the original practice so that an alternative practice deserves a fair hearing. If the new practice looks better we shift to it; we need not (and in this example cannot) argue to it on the basis of the old perceptual practice.

In this example, in fact, it is plausibly maintained that the new practice can be argued to be better than the old practice *independently of any specific problems with the old practice.* For the new practice (involving only judgements about angle from the horizon, rather than about absolute motion) is less committal than the old practice and yet serves all our purposes just as well, so that it has advantages of economy over the old practice. If one accepts this, then the bearing on platonism is apparent: for the fictionalist's judgements about naturalness are less committal than the platonist's plausibility judgements, so one should shift to the fictionalist's practice unless one can argue that there are respects in which the platonist's practice serves our purposes better. *And the only way to argue that would be to appeal to considerations independent of initial plausibility, such as indispensability arguments.*[7]

---

[7] One can't very well avoid this conclusion by simply saying that the platonist practice has advantages over the fictionalist practice, in that it leads us to truths about mathematical entities; that would be like saying that the perceptual practice that licenses beliefs about absolute motion has advantages over the new one, in that it leads us to the facts about absolute motion. The point is that the alleged 'facts' about absolute motion need to be shown to be important; similarly for the alleged 'facts' about mathematical objects. Maybe they can be shown to be important, for instance by indispensability arguments: here I am only arguing that that is where the interesting issues lie.

The platonist might want to protest that what undermined the old perceptual practice in this example isn't just the existence of a new perceptual practice that is less committal, but also a critique of the old perceptual practice. That is, part of what was involved in the development of the new perceptual practice was the development of and/or emphasis on the concept of relative motion; *this then led to the realization that it is only relative motion that can be reliably detected by direct visual observation*, and it is that critique of the old observational practice, not *just* the development of the alternative practice and it's being less committal, that led to the undermining of the old practice. I have to agree that this response to the example makes a certain amount of sense. But it should be noted that a platonist who responds in this way implicitly admits that the platonist practice of making initial plausibility judgements about mathematical entities could be undermined if it were subjected to an analogous critique: that is, the platonist practice would be undermined if it could be argued that even on the supposition that there are mathematical facts, there is no way in which our plausibility judgements could be expected to reliably reflect them. Arguments of roughly this form *have* been suggested by various people (see section 4 of this introduction, and section 2 of essay 7). So our discussion leads, if not to the strong conclusion of the last paragraph, then at least to this weaker conclusion: even if we grant that initial plausibility judgements in mathematics are analogous to perceptual judgements, still arguments of the sort just alluded to might undermine them, and lead to their replacement by the less committal judgements of naturalness that even the fictionalist accepts.

There is another, simpler, route to the conclusion I have drawn – indeed, to the stronger conclusion of two paragraphs back, rather than the more concessive one of the last paragraph. This argument too will be made by analogy. Consider an ontological dispute about, not mathematical entities, but another somewhat controversial sort of entity: regions of space. Let us use the term 'substantivalist' for a person who believes that talk of regions of space is true taken at face value, that is, without reconstrual in terms of physical objects. Now, if one is a substantivalist, it will make perfectly good sense to say that one makes perceptual judgements about regions of space – for instance, it makes sense to say that one perceives that the region of space in front of one is empty of visible objects. So claims about space can have *perceptual* initial plausibility, and that is the kind of initial plausibility whose epistemological value is presumably least controversial. *Even so*, the claim that one can have perceptual knowledge of physical space does nothing to undermine a challenge to substantivalism (and to the practice of making substantivalist perceptual judgements) on the basis of ontological economy.[8]

The reason, as before, is that observational reports are always theory-laden; and this example (like the previous one) makes clear that if the theory with which they are laden is challenged by someone with an alternative theory, one's practice of making observational reports in the way that one does needs defence against that challenger. Faced with a relationalist who thinks that there is no need to postulate physical space (as anything other than a manner of speaking about ordinary objects), a believer in physical space needs to reply on the relationalist's own terms: to claim that one has perceptual knowledge that there are regions of physical space devoid of visible objects would in *that* context carry no epistemological weight. In essay 6 below I attempt to meet the relationalist on his or her own terms, with two types of indispensability arguments. My present point is simply that the platonist ought to respond to the fictionalist in the same manner, and that until he or she does so there is no reason to persist in regarding mathematical claims as initially plausible (rather than simply as being natural to accept for diverse reasons of the sorts I have described).

3 Indispensability Arguments and Inference to the Best Explanation

As I said early in the previous section, there *are prima facie* difficulties in maintaining a fictionalist position about mathematics: at least, there are for anyone who takes indispensability arguments seriously, as I do. An indispensability argument is an argument that we should believe a certain claim (for instance, a claim asserting the existence of a certain kind of entity) because doing so is indispensable for certain purposes (which the argument then details). In this section I will focus on one special kind of indispensability argument: one involving indispensability *for explanations*. (There is some discussion of other kinds of indispensability argument in several of the essays in this volume, especially essays 3 and 7.) To rely on this special kind of indispensability argument is to rely on a principle of 'inference to the best explanation'. Some such principle seems to underlie much of our knowledge of the physical world.

---

[8] I don't say that the claim that space-time regions are perceivable is of no epistemological importance: my claim is rather that it has importance only in a different kind of epistemological context than that which is here in question. In this different context, the existence of unproblematic connections between perceptual judgements and perceptual facts, not just the initial plausibility of perceptual judgements, are important. I will return to this in section 4.

More fully, suppose (a) that we have certain beliefs, beliefs about 'the phenomena', which we are unwilling to give up; (b) that this class of 'phenomena' that we believe in is large and complex; (c) that we have a pretty good explanation of these phenomena (in the sense of, a relatively simple non-*ad hoc* body of principles from which they follow); and (d) one of the assumptions that appears in this explanation is claim S, and we are pretty sure that no explanation of the phenomena that does without claim S is possible. The idea of 'inference to the best explanation' is that under these circumstances we have a strong reason to believe claim S. (If this seems vacuous, I should add that we do not accept as explanations claims of the form 'The phenomena are as they would be if explanation E were correct': as-if claims which ride piggyback on genuine explanations are not themselves to be construed as explanations (at least, non-*ad hoc* ones), for the purpose of understanding this principle. Given this, the principle, though of course vaguely formulated, is non-vacuous: it precludes us from accepting a large and complex set of phenomena as brute when there is a decent explanation of these phenomena in the offing.)

It seems to me that most of us accept the principle of 'inference to the best explanation', in the sense that this principle (or something pretty close to it) governs our ordinary inductive methodology.[9] Clearly something like inference to the best explanation is at work in us in giving rise to observational beliefs, that is, beliefs that could be independently checked by observation: we come to believe that a pipe behind the wall is leaking, since this best explains the stains on the wallpaper, the warped floor-boards, etc. The same principle seems to be involved when we arrive at beliefs about unobservable entities, or non-observational beliefs about observables. The principle of inference to the best explanation makes no discrimination among these three cases: if a belief plays an ineliminable role in explanations of our observations, then other things being equal we should believe it, regardless of whether that belief is itself observational, and regardless of whether the entities it is about are observable. That I think is the methodology we (nearly) all employ, and I think it would be unwise to change it. The fact that the principle does not discriminate over whether the explanation is observational (or whether it postulates unobservable entities) stands up well to reflection: intuitively, the observational nature of the explanation should make no difference in an inference to the best explanation. After all, in *any* case where we rely on inference to the best explanation, our belief goes beyond what we have observed; the fact that one belief *could* be fairly directly tested by observation while the other *couldn't*

---

[9] For a contrary view, see van Fraassen (1980).

seems to have no relevance to their evidential status *when such an independent test has not been made*. (When the independent test *has* been made – when the leak behind the wall has been directly observed – then we need no longer rely on inference to the best explanation. When we do rely on inference to the best explanation, our beliefs go beyond the observations we have made, and my point is that the difference with respect to *possible* observations that *haven't* been made is irrelevant to our *actual* evidential situation.)

To be sure, it would be possible to introduce a restricted form of inference to the best explanation that licenses only observational beliefs; but, firstly, this would be entirely *ad hoc* and unmotivated,[10] and, secondly, such a restriction would cripple our beliefs about observables. Many observational beliefs (e.g., the beliefs that the Los Alamos scientists had about what would happen when they made the first atomic test) seem to depend on beliefs about unobservables; they are not obtainable directly from other claims about observables, without detour through the unobservables. (For instance, they aren't obtainable from past observations by any straightforward sort of enumerative induction, since they concern observable situations very different from any hitherto encountered.) I don't deny that one could formulate a principle that would allow belief in the observable (though hitherto unobserved) consequences of inferences to not observationally checkable explanations, without believing in those explanations themselves, but such a position seems to me even more *ad hoc* than the simple restriction to observationally checkable explanations. (Also, the fact that it would deprive us of believable explanations of the things we believe – deprive us of a simple and unified body of beliefs from which many of the rest of our beliefs follow – seems to me to make it intrinsically unattractive.)

But if our belief in electrons and neutrinos is justified by something like inference to the best explanation, isn't our belief in numbers and functions and other mathematical entities equally justified by the same methodology? After all, the theories that we use in explaining various facts about the physical world not only involve a commitment to electrons and neutrinos, they involve a commitment to numbers and functions and the like. (For instance, they say things like 'there is a bilinear differentiable function, the electromagnetic field function, that

---

[10]   If one is going to make an unmotivated restriction that we use inference to the best explanation only to arrive at observational beliefs (beliefs directly checkable in principle by human beings, wherever and whenever located), why not the further unmotivated restriction that we use it only to arrive at beliefs directly checkable *by me*, or directly checkable in principle *by beings located in this galaxy during the check*, or directly checkable *in an experiment cheap enough for the government to fund*?

assigns a number to each triple consisting of a space-time point and two vectors located at that point, and it obeys Maxwell's equations and the Lorentz force law.') I think that this sort of argument for the existence of mathematical entities (the Quine–Putnam argument, I'll call it) is an extremely powerful one, at least *prima facie*. It should be noted that the Quine–Putnam argument is not merely that just as there are good explanations in which the postulation of unobservables is essential, so too are there good explanations in which the postulation of mathematical entities is essential, so that if inference to the best explanation licenses one it licenses the other. The argument is stronger, in that it says that the very same explanations in which the postulation of unobservables is essential are explanations in which the postulation of mathematical entities is essential: mathematics enters essentially into our theory of (say) electrons. There seems to be no possibility of accepting electrons on the basis of inference to the best explanation, but not accepting mathematical entities on that basis, by saying that the explanations involving the latter are weaker than the explanations involving the former: for the very same explanations are involved in both cases. This fact makes arguments for the indispensability of mathematical entities *in explanations of the physical world* seem in some ways more compelling to a scientific realist than other indispensability arguments. (Not that other indispensability arguments may not be compelling too.)

Still, two points need to be made about justifying belief in mathematics by its apparent indispensability in science (or elsewhere). Together, they make the ultimate force of the Quine–Putnam argument hard to ascertain.

The first point is that any such 'indispensability to science' justification of mathematics can be undercut if we can show that there are equally good theories and explanations that don't involve commitment to numbers and functions and the like. I believe that such an undercutting of the justification is possible (but that no analogous undercutting is possible in the case of our justification for believing in electrons and neutrinos and the like). I originally made a case for this in my book *Science without Numbers*.

At present of course we do not know in detail how to eliminate mathematical entities from every scientific explanation we accept; consequently, I think that our inductive methodology does at present give us some justification for believing in mathematical entities. But this brings me to my second point, which is that justification is not an all or nothing affair. The belief in mathematical entities raises some problems which I and many others believe to be fairly serious. (I will briefly discuss two of those problems in the next section.) These puzzles provide reasons against the belief in mathematical entities, and to put it very crudely, what we must do is weigh the reasons for and the reasons against in deciding what to believe. Less crudely, what we must do is make a bet on how best to achieve a satisfactory overall view of the place of mathematics in the world. I do not declare that it is misguided to try to solve the puzzles that many people have found in platonism; on the contrary, much interesting work is being done in that direction, and I hope that a large segment of the philosophical community continues to pursue that line of research. Let a hundred flowers bloom. But my tentative bet is that we would do better to try to show that the explanatory role of mathematical entities is not what it superficially appears to be; and the most convincing way to do that would be to show that there are some fairly general strategies that can be employed to purge theories of all reference to mathematical entities.[11] In any case, I think it is an idea that needs to be pursued much further than the philosophical community has pursued it so far.

I think, indeed, that showing how to eliminate mathematical entities from explanations would be attractive for reasons not wholly dependent on anti-platonism. For *even on the assumption that mathematical entities exist*, there is a *prima facie* oddity in thinking that they enter crucially into explanations of what is going on in the non-platonic realm of matter. It seems to me that the most satisfying explanations are usually 'intrinsic' ones that don't invoke entities that are causally irrelevant to what is being explained.[12] 'Extrinsic' explanations are acceptable (as when we explain the behaviour of a non-human or non-English speaker by reference to English sentences that he or she believes or desires), but it is natural to think that for any good extrinsic explanation there is an intrinsic explanation that underlies it. This principle seems plausible independently of anti-platonist scruples (it wouldn't help to refer to inscriptions of English sentences instead of sentence types of English in our explanation of the non-human's or foreigner's behaviour); but it requires that there are mathematical entity free explanations underlying the platonistic ones, since mathematical entities (according to the views

---

[11]   The result of such a purging would be a theory that a physicist would probably regard as merely a rewritten version of the original theory; just as a physicist would probably regard Newtonian physics formulated without talk of absolute rest as merely a rewritten version of Newtonian physics with absolute rest. But for a philosopher, there is an advantage in formulating Newtonian physics without absolute rest, and so too for formulating it without mathematical entities. What counts as two formulations of the same theory is a context-relative matter.

[12]   This is over-simplified: a more careful formulation is needed to handle non-local quantum phenomena; also, independent of quantum considerations, a suitable kind of spatio-temporal connection may be enough for 'intrinsicness' even when the causal condition fails. (The alternative condition in terms of spatio-temporal connection might be enough to handle the quantum phenomena too.)

of those who believe in them) don't enter into causal interactions with the material world (in that for instance there is no interchange of energy-momentum between them and the material world).[13] I regard the acceptance of an extrinsic explanation as ultimate as at least slightly odd.

Whether or not one takes this *prima facie* oddity seriously as a motivation for showing that mathematical entities are in principle eliminable from physical theory, it does point up an important fact: the role of mathematical entities, in our explanations of the physical world, is very different from the role of physical entities in the same explanations. For the most part, the role of physical entities in those explanations is causal: they are assumed to be causal agents with a causal role in producing the phenomena to be explained. Since mathematical entities are assumed to be acausal, their explanatory role (or roles) must be somehow different.

It may be thought that the difference in the explanatory role of mathematical entities and physical entities is enough to motivate a restriction of inference to the best explanation to the latter. The position would be (1) that we should literally believe in the existence of electrons and their properties as postulated in our physical theories, since there are good explanations in which they are assumed causally relevant, and

---

[13]   Here I exempt the views of Popper (1972), who seems to hold that mathematical entities can causally affect our immaterial minds, and thereby indirectly affect our bodies by way of our pineal glands. (Gödel is sometimes taken to hold this too, on the basis of some much-quoted remarks in his 1944 and 1947.)

Note also that weakening the notion of intrinsicness in accordance with the previous footnote is unlikely to help, since, among other things, mathematical entities are usually taken to have no spatio-temporal location. Again there is an exception: Maddy (1980) holds that at least some mathematical entities have spatial location. ({Reagan}, {{Reagan}, Reagan}, and so forth, all have the same location, namely that of Reagan himself; similarly, {Reagan, Carter}, {Reagan, {Carter}}, etc. all have the same location as does the aggregate of Reagan and Carter. The number 3 is a property of all three-membered sets; presumably each region of space is the location of a three-membered set (say, the set of its left third, its middle third and its right third), so that 3 is a property that is instantiated everywhere. I don't know if the number 3.782 and the exponential function have spatial location on her view, or if she views them as properties of things that have locations; but I suspect that if they do have or apply to things that have location, they are located or instantiated everywhere.) Maddy's view, unlike Popper's, does not strike me as at all silly. Rather, it is a perfectly sensible convention about how to talk about mathematical entities. I doubt, though, that it is enough to resolve worries about the explanatory legitimacy of the use of mathematical entities in explanations, especially the more complex ones typical of advanced physics. I should add that it was not her intention to deal with *these* worries; she was concerned rather with the epistemological issues that we will deal with in the second half of the next section. I suspect that the doubts I have raised about how helpful it is to assign sets spatial location apply in that context too, but there the issues are in some ways more complicated.

there is no obvious prospect of eliminating them from these explanations; but (2) that we shouldn't literally believe in mathematical entities, since there are no good explanations *in which they are assumed to be causally relevant*, despite the fact that there is no way of giving explanations that avoids postulating them in an *acausal* role. The problem is that if one takes this line, then the properties of electrons that one literally believes in can't include any properties that require mathematical entities for their expression; so if mathematics is not eliminable or close to eliminable, there are going to be very serious limitations on stating explanations in terms of electrons without going beyond what one believes. Perhaps one can maintain a belief in electrons and a belief in those of their properties which are describable without mathematics, on *something like* inference to the best explanation grounds, without a belief in the explanations one gives; but it seems to me a very delicate position to maintain.[14] Consequently I am inclined to think that unless a very substantial amount of explanation involving electrons can be given in a mathematical entity free fashion, the prospects for maintaining realism about electrons without maintaining platonism are dim.[15]

4 Problems with Platonism

I do not intend to discuss in detail the various problems that many have felt to arise in the platonist position. It is, though, worth commenting briefly on two of them.

A. 'What Numbers Could Not Be'

A noteworthy feature of mathematics is that there is a tremendous amount of arbitrariness as to the identification of different types of mathematical objects. The most famous example of this is the one highlighted by Paul Benacerraf (1965): if one wants to identify natural numbers with sets, it seems rather arbitrary which sets one picks. But of course this example is just the tip of the iceberg: just as there is no uniquely natural set-theoretic explication of natural numbers, there is no uniquely natural set-theoretic explication of real numbers (for instance, one can use Dedekind cuts or equivalence classes of Cauchy

---

[14] See Cartwright (1983) for an attempt to maintain it.

[15] I like to set my goals high – I like to set myself the task of entirely doing without mathematical entities in physical theory. It may prove in the end that this task cannot be carried out, but that substantial parts of it can be carried out: that substantial bodies of explanation can be given in a mathematical entity free way, even if not full physical theory. If that were to turn out the case, I think that the position contemplated in this paragraph would need to be seriously considered.

sequences); or of ordered pairs; or of the tensor product of two vector spaces; or of the tangent vectors at a point on a manifold; and so on *ad infinitum*. It seems absurd to suggest that in each such case there is a fact of the matter as to which sets the mathematical objects in question are. Indeed, the problem is deeper than this: for there is also an arbitrariness in thinking that it is sets that are to be regarded as the basic objects. As Benacerraf mentions, it is possible to take ordinal numbers as basic and define sets in terms of them. Also, we could take functions as basic, and define sets as functions of a certain sort (say, identity functions); or we could take relations as our basic objects, and define sets as relations of adicity 1. The choice between regarding relations as sets of a certain sort and regarding sets as relations of a certain sort seems no more factual than the choice between regarding natural numbers as Zermelo sets and regarding them as von Neumann sets; or between regarding real numbers as Dedekind cuts or as equivalence classes of Cauchy sequences; or between regarding the tensor product of the vector spaces V_1 and V_2 as a space whose elements are equivalence classes of formal sums of pairs of vectors, and regarding it as a space whose elements are the linear functionals on the bilinear forms on the direct sum of V_1 and V_2.

But what exactly is the significance of the arbitrariness in these choices? This is of course quite controversial. In the case of the identifications of natural numbers with sets, some philosophers (for instance, Steiner 1975) have taken the position that the arbitrariness of one identification over the others shows that all such identifications are false: numbers and sets are simply distinct sorts of entities. Perhaps this position has some plausibility in this case (and in the case of the identifications of real numbers with sets, and in a few other special cases); but surely as a general position it defies credibility. Are we to say, for instance, that a topological space is distinct from any set-theoretic construct, since it could be taken to be either a pair of a set and the set of its open subsets, or of the set and its set of closed subsets, or of the set and its closure operator? And are we to say the same for each other mathematical structure, even the arcane and specialized ones? Perhaps saying this is *slightly* more natural than saying that topological spaces *definitely are* set-theoretic constructs of some particular sort. But I think that by far the most natural conclusion is that topological spaces, and numbers, and ordered pairs and functions are neither definitely sets nor definitely not sets: there is no fact of the matter about whether they are sets or not, in addition to there being no fact of the matter as to which sets they are if they are sets.

Now, supposing that the arbitrariness in identifying one type of mathematical object with another is as pervasive as I have been saying it is, how are we to explain this? I think that by far the most natural explanation of the pervasive arbitrariness (an explanation persuasively developed in Wagner 1982) is the fictionalist one: we have a good story about natural numbers, another good story about sets, and so forth; and in these stories it is completely unimportant whether one identifies numbers with sets, and unimportant which sets one identifies them with if one does want an identification. As Wagner says, to offer a particular identification of numbers with sets, or to deny all such identifications, is comparable to saying what happened to Little Red Riding Hood in her adult life: it is hardly surprising that there is no matter of correctness here, if one views mathematics as fiction. (Surprisingly, Wagner confines his fictionalism to arithmetic. But it would seem that the argument applies to mathematics rather generally, if the remarks above about the pervasiveness of arbitrary choice in the identification of mathematical entities are correct.)

But can a platonist also find a way to grant that there is no fact of the matter as to the truth of cross-theory identities in mathematics? There seems to be an obstacle to so doing: after all, if the number 2 is a definite object, and the set $\{\varnothing, \{\varnothing\}\}$ is a definite object, doesn't there have to be a definite question as to whether the former is the latter? I would not say that there is no possible platonist answer to this question; but I doubt that there is any that is nearly so plausible as the fictionalist answer.

One initially attractive platonist view, suggested by Benacerraf, is a 'structuralist' view according to which there are, literally speaking, no such things as numbers, but there are mathematical structures that are ω-sequences, and talk of numbers is just a convenient way of talking about ω-sequences. (See also White 1974 for a slight variant.) The trouble with that view (as Kitcher 1978 notes) is that if numbers aren't acceptable entities because of the arbitrariness of whether and how to identify them with sets, ω-sequences are no more acceptable. Indeed, even if one agrees to identify them (somewhat arbitrarily) with relations and to identify relations (somewhat arbitrarily) with sets of ordered pairs, the arbitrariness of the identification of ordered pairs with sets is at least as great as the arbitrariness in the identification of numbers with sets.

There is a variant of the Benacerraf–White view which I think immune to this objection: it is one I mentioned some years ago (Field 1974). The basic idea is that talk of numbers, and of other mathematical objects including ω-sequences and sets, has a considerable degree of vagueness or referential indeterminacy. We have to accept that indeterminacy as a fact of life: there is no possibility of eliminating it by shifting from talk of numbers to talk of ω-sequences. (I used ω-sequences in sketching a metalinguistic account of truth for sentences containing the numerals, but it should have been clear from the rest of the paper that the metalanguage was not supposed to be any more referentially determinate than the object language.) There are many questions that could be raised about this view; but I will not pursue them, for I think that when one compares this view with the fictionalist treatment that Wagner recommends, the Wagner view looks much more attractive. So I am inclined to share Wagner's view that the considerations of cross-theory identities do provide substantial motivation for fictionalism.

Crispin Wright (1983) has recently argued that Benacerraf's paper raises no special problems for mathematics. He writes, 'Benacerraf complains that nothing in our use of numerical singular terms is sufficient to determine which, if any, classes they stand for. But the same is evidently true of singular terms standing for classes themselves' (p. 125). He goes on to remind us that Quine drew a similar conclusion not only for terms purportedly referring to mathematical entities, but for terms purportedly referring to ordinary physical objects like rabbits. And he concludes by saying

But Benacerraf's anti-platonist [sic] conclusion needs to appeal to a tacit supplementary premise: that where standard uses and explanations are insufficient to determine uniquely the putative references of an apparently referential expression, that expression is not genuinely referential . . . We ought to draw Benacerraf's anti-platonist conclusion only if, confronting similar indeterminacies with respect to the language of classes itself, or our talk of rabbits, we are prepared to infer similarly that singular reference to classes, or to rabbits, is illusory. (p. 127)

It seems to me that Wright's remarks are simply not addressed to Benacerraf's argument. They are addressed to a different anti-platonist argument, one with some appeal: the argument that we ought to be suspicious about numbers because if we assume that they exist, it seems impossible to explain how we can refer to them or have beliefs about them. (Probably the most persuasive statement of that different argument is in Hodes 1984b.) According to that argument, our mathematical practice is sufficient to ensure that the entities to which we apply our word 'number' form an ω-sequence of distinct objects, under the relation we call '$<$'; but that is the most that our use of these number-theoretic words determines. (Indeed, maybe our usage doesn't determine even this much: maybe it determines only that the entities to which 'number' applies form a sequence that satisfies our best axiomatic first order theory of ω-sequences; that is, all that is determined is that they form a possibly non-standard model of such a theory.) Presumably anyone who advocates this sort of position (with or without the parenthetical extension) in the case of numbers will advocate the same conclusion in the case of sets: Wright's point that an anti-platonist who deploys this argument against numbers should deploy it against sets too is certainly correct. Wright's claim that the argument has equal force against rabbits is more controversial: certainly advocates of causal theories of reference would claim that causal considerations do much to constrain the reference of 'rabbit', and it is precisely the fact that such causal considerations seem inapplicable in the case of numbers and sets that makes those so much more problematic.

But suppose that Wright is correct. Suppose, as he says, that we don't need to assume that 'where standard uses and explanations are insufficient to determine uniquely the putative reference of an apparently referential expression, that expression is not genuinely referential'. Presumably if we are to draw this conclusion it is because our notion of reference is disquotational. On a disquotational view of reference, we are entitled to keep the disquotation schemata

> If b exists then 'b' refers to b

and

> 'F' applies to Fs and to nothing else

independent of any 'theory of reference'. These schemata allow us to assert that 'set' applies to sets, that 'rabbit' applies to rabbits, that 'number' applies to numbers, that '2' refers to 2. But the schemata do not allow us to assert that 'number' does, or that it does not, apply to sets; nor do they allow us to assert, or to deny, that if '2' refers to a set then it refers to $\{\varnothing, \{\varnothing\}\}$. To assert or deny these things, we would need not only disquotational reference but also a claim about the identity or non-identity of mathematical entities in different theories. And it was of course the issue of identities and non-identities, not the issue of reference, with which Benacerraf's argument was concerned; Wright's points are irrelevant to the argument that Benacerraf actually gave.

I do not deny that there is a counter to Benacerraf with some similarity to Wright's that needs addressing: the counter is that (not the problem of reference but) the pervasiveness of arbitrariness about identifications is a feature of the non-mathematical realm as well as of the mathematical. (That claim is, I believe, the most interesting feature of Putnam's critique of 'metaphysical realism', in his 1981 and elsewhere.) A typical example: it is sometimes said that it is arbitrary whether we take a point of space to be simply a region of minimal size, with zero (or infinitesimal) volume; or instead say that a point is a convergent set of smaller and smaller regions, each region in the set having non-zero (and non-infinitesimal) volume. This example, of course, will not impress the anti-platonist: if one rejects sets, then *literally* speaking there are no convergent sets of regions; moreover, the task of making do with only regions of non-zero (and non-infinitesimal) volume in developing a non-platonistic physics is highly non-trivial. Whether other examples are immune to this sort of reply (and to various other replies) is not an issue I can pursue here, but my own view is that we do not get the same kind of pervasive arbitrariness at the purely physical level that we do at the mathematical.

B. Knowledge of Mathematical Entities

Perhaps the most widely discussed challenge to the platonist position is epistemological. Here the *locus classicus* is again a paper by Benacerraf (1973). Benacerraf's formulation of the challenge relied on a causal theory of knowledge which almost no one believes anymore; but I think that he was on to a much deeper difficulty for platonism.

Very roughly, Benacerraf's challenge can be put like this: if there are mathematical entities of the sort that the platonist believes in (mind- and language-independent, having no spatio-temporal location, unable to enter into physical interactions with us or anything we can observe) then there seems to be a difficulty in seeing how we could ever know that they exist, or know anything about them; the platonist needs to explain how such knowledge is possible, and no answer is evident except one that posits mysterious powers of access to the platonic realm. (Note that it is not *just* the acausal character of the mathematical entities that gives rise to the apparent problem; rather it is a combination of characteristics that collectively make access to the entities seem mysteri-ous.) It may seem that if the previous section is correct then we have an answer to this epistemological question: we know about mathematical entities because theories that postulate them and attribute specific properties to them are indispensable in our various theories – for instance, in our physical theories. Of course, this assumes that mathematical entities *are* indispensable (to physical theory or to some other important body of extra-mathematical belief), and this is an assumption that a fictionalist (of my sort) would question. But I think that there is a more fundamental problem with this answer to Benacerraf: I think that we can formulate his challenge more carefully, so as to make indispensability considerations of questionable relevance in answering it.

The way to understand Benacerraf's challenge, I think, is not as a challenge to our ability to *justify* our mathematical beliefs, but as a challenge to our ability to *explain the reliability* of these beliefs. We start out by assuming the existence of mathematical entities that obey the standard mathematical theories; we grant also that there may be positive reasons for believing in those entities. These positive reasons might involve only initial plausibility, for those who are unconvinced of my treatment of initial plausibility in section 2. Alternatively, the positive reasons might be that the postulation of these entities appears to be indispensable for some important purposes. But Benacerraf's challenge – or at least, the challenge which his paper suggests to me – is to provide an account of the mechanisms that explain how our beliefs about these remote entities can so well reflect the facts about them. The idea is that *if it appears in principle impossible to explain this*, then that tends to *undermine* the belief in mathematical entities, *despite* whatever reason we might have for believing in them. Of course, the reasons for believing in mathematical entities (in particular, the indispensability arguments) still need to be addressed, but the role of the Benacerrafian challenge (as I see it) is to raise the cost of thinking that the postulation of mathematical entities is a proper solution, and to thereby increase the motivation for showing that mathematics is not really indispensable after all.

I am aware, of course, that this sketch of the form I take Benacerraf's challenge to have is highly schematic. To fill it out, one would have to do four things. First, one would have to formulate more clearly the claim that our mathematical beliefs are 'reliable' or 'reflect the mathematical facts'. In essay 7 below I argue that in doing this we need not rely on any notion of fact, or even on any notion of truth beyond a thoroughly disquotational one: the claim is simply that the following schema

> If mathematicians accept 'p' then p

(and a partial but hard to state converse of it) holds in nearly all instances, when 'p' is replaced by a mathematical sentence. The second thing one would need to do is argue that a platonist needs to accept this 'reliability' claim; I think, though, that the platonist's need to do this is beyond serious question. The third thing one must do is argue that a platonist must not only *accept* the reliability, but must commit himself or herself to the possibility of *explaining* it. The idea is that the correlation between mathematicians' belief states and the mathematical facts postulated in the above schema (and its partial converse) is so striking as to demand explanation; it is not the sort of fact that is comfortably taken as brute. (The platonist can legitimately postulate brute facts about mathematical entities themselves, for instance, basic laws of set theory; and even certain kinds of brute facts about the relations between mathematical entities and physical entities, for instance that every physical entity is a member of some set. But special 'reliability relations' between the mathematical realm and the belief states of mathematicians seem altogether too much to swallow. It is rather as if someone claimed that his or her belief states about the daily happenings in a remote village in Nepal were nearly all disquotationally true, despite the absence of any mechanism to explain the correlation between those belief states and the happenings in the village. Surely we should accept this only as a very last resort.) Fourth and finally, to make it believable that the Benacerrafian challenge is insurmountable, one would have to argue that it is impossible to explain the reliability claim in question: one would have to argue that various facts about how the platonist conceives of mathematical objects collectively rule out all possibility of finding any such explanation. (The relevant facts about how the platonist conceives of mathematical objects include their mind-independence and language-independence; the fact that they bear no spatio-temporal relations to us; the fact that they do not undergo any physical interactions (exchanges of energy-momentum and the like) with us or anything we can observe; etc.) Like Benacerraf, I refrain from making any sweeping assertion about the impossibility of the required explanation. However, I am not at all optimistic about the prospects of providing it.

Several points about this are worth making here. First, it seems to me that something like the problem here under discussion has been a main motivation for various versions of what I've called 'mathematical idealism': that is, for various views according to which mathematical entities are some kind of 'mental constructions' (or 'constructions out of our linguistic practices'). Advocates of such views assume, I think, that it would not be hard to explain why our beliefs are reliably correlated with facts that we ourselves have constructed. Whether or not they are right about this is hard to say: talk of mathematical entities as 'constructed by' the mind (or by our linguistic practices) strikes me as so obscure that until it is explained, no answer is possible. As I remarked earlier on, it may be best to interpret such talk of 'constructions' as simply a picturesque way of saying that mathematical talk should be interpreted along fictionalist lines.[16]

---

[16] If one does not so construe them as restatements of fictionalism, it seems to me that there are two dangers to which they *may* be liable. (Whether they really are so liable depends on how talk of 'constructions' is to be understood.) The first danger is that one may not be able to make sense of all of classical mathematics if one tries to impose an idealist construal of it. (It was an idealist view of mathematics that led Brouwer and Heyting to intuitionism – see for instance Heyting 1956.) The second danger is that on a limited idealist view, one that views mathematical entities as some sort of human construction but makes no such claim about the physical world, the application of mathematics to the physical world may turn out to be a mystery. The danger, in other words, is that in order to explain the applicability of mind-dependent mathematical entities to the physical world, the idealist about mathematics may have to become a full-blown idealist, and hold that even things like electrons and dinosaurs are somehow 'human constructions'. If this danger were indeed realized, I would regard that as a *reductio ad absurdum* of the idea that mathematical objects were human constructions. I do not want to assert that it is impossible to develop a mathematical idealism that avoids both dangers and is genuinely distinct from fictionalism and succeeds in solving the epistemological problem (and various other problems) that the idealist finds with the platonist position; but I must confess to having little idea how it might be done.

A second point about the Benacerraf problem as I have reconstructed it: it is sometimes said that there is a need to explain the reliability of our beliefs about entities of a certain type *when the facts those beliefs report are contingent*; but that in the case of mathematical entities the facts in question hold necessarily, and this makes the task of explaining the reliability of our beliefs trivial or unnecessary. I respond to such views at some length in essay 7 below, and will say no more about them here.

Third, the fact that some mathematical claims may seem initially plausible is no help in responding to the version of Benacerraf's problem that I have sketched. Claims of initial plausibility are of some help to the platonist in answering questions about justification; as I argued in section 2, they are helpful in answering questions about the justification of particular mathematical beliefs, at least relative to a certain practice of making plausibility judgements. (I also argued there that this relative justification did not give them any special authority in contexts where that practice is itself questioned; but my present point is independent of this.) But to give them a justificatory role does nothing to explain the reliability of this class of judgements. Someone *could* try to explain the reliability of these initially plausible mathematical judgements by saying that we have a special faculty of mathematical intuition that allows us direct access to the mathematical realm. I take it though that this is a desperate move, rather akin to the move of postulating a special faculty of intuition that allows the character three paragraphs back direct access to the events in Nepal.

The fourth point I want to make is more concessive to the platonist, or at least, to the platonist who bases his or her platonism on some sort of indispensability argument – especially one who stresses the indispensability of mathematics in application to the physical world. For one can try to invoke indispensability considerations not simply in the context of justification, but in the context of explaining reliability. One could argue, for instance, that if mathematics is indispensable to the laws of empirical science, then *if the mathematical facts were different, different empirical consequences could be derived from the same laws of (mathematized) physics.*[17] So, it could be said, mathematical facts make an empirical difference, and maybe this would enable the application-based platonist to argue that our observations of the empirical

---

[17] This does not conflict with the conservativeness of mathematics: that has nothing to say about what happens when you apply mathematics to platonistic physical laws. (Also, it's only the actual mathematical facts that a platonist must admit are conservative. It isn't obvious that the platonist should have to agree that mathematics would still be conservative if the mathematical facts were different.)

consequences of physical law are enough to explain the reliability of our mathematical beliefs. An advocate of this indispensability line might even argue that initial plausibility judgements play an important role in explaining the reliability of our mathematical beliefs: the idea would be that evolutionary pressures (biological and/or cultural ones) have led us to find initially plausible those mathematical claims which are empirically indispensable, and that this gives all the explanation of the correlation between our judgements and the mathematical facts that we should want.

I'm suspicious about this line of response (with or without the extension that encompasses initial plausibility judgements) to the Benacerrafian challenge; but my most general worries about it involve some large issues, and I think it better not to attempt to raise them here. (I suspect that it is impossible to deal adequately with these general worries separately from some of the issues briefly touched on at the end of section 3, about the differences between the explanatory roles of mathematical entities and of physical entities.) But I will raise two more specific doubts about the prospects for dismissing the Benacerrafian challenge in this way. The first specific worry is that the amount of mathematics that gets applied in empirical science (or indeed, in metalogic and in other areas where mathematics gets applied) is relatively small. This means that only the reliability of a small part of our mathematical beliefs could be directly explained by the proposal of the previous paragraph. To be sure, one could try to use the reliability of our beliefs in this relatively small part of mathematics to 'bootstrap up' to the reliability of larger parts, by hypothetico deductive inference within mathematics: see the discussion of the quotations from Gödel in essay 2.[18] But I think that there is substantial room to doubt that such inferences are all that powerful: too many different answers to questions about, say, large cardinals or the continuum hypothesis or even the axiom of choice work well enough at giving us the lower level mathematics needed in science and elsewhere. (One could of course just admit that we are and always will be ignorant of the mathematical facts about the continuum hypothesis and the axiom of choice and even the small large cardinals, but I don't think that this is an attitude many mathematicians would find attractive. In section 8C of essay 7, I describe an alternative and I think more appealing viewpoint toward these axioms, one which allows there to be reasons for preferring some axioms to others while denying that the choice is a matter of truth value about which we might be mistaken.)

---

[18] For a more thorough elaboration of a position like Gödel's, see Maddy (1988).

My second specific reason for doubting the adequacy of the reply of two paragraphs back to the reliability worry is really an extension of the first. The first worry began with the fact that the amount of mathematics employed in empirical science (and elsewhere, e.g. in metalogic) is relatively small. This is so quite independently of any partial successes of the programme of nominalizing science (and metalogic, etc.). But if, which I take as true, the partial successes of the nominalization programme have been substantial, this very much weakens the case for the reliability of the mathematical beliefs that we apparently need in those cases where the nominalization programme has not been carried out. Suppose for instance that we could nominalize everything but quantum theory. If this were so (and if my earlier critique of autonomous platonism is correct) then the entire weight of our belief in mathematical entities would rest on quantum theory. Is it really believable that an adequate account of the reliability of our mathematical beliefs could be made on this basis?

Of course, in actual fact quantum mechanics is not the only thing that has so far resisted nominalization, but the general point is clear: the more the partial successes of the nominalization programme, the more the difficulties for the attempt to respond to the Benacerrafian problem on indispensabilist lines, and therefore the more the motivation to try to complete the nominalization programme so that we can maintain a fictionalist view on which the Benacerraf problem does not arise.

PART TWO

5 Logical Implication

What should a fictionalist say about such metalogical notions as logical implication and logical consistency? The standard definitions of logical implication and logical consistency (due to Tarski 1956) are in terms of models. Suppose that Γ is a set of sentences, and B is a sentence; then

(i) Γ logically implies B if and only if B is true in every model in which all members of Γ are true;

(ii) Γ is logically consistent if and only if there is at least one model in which all members of Γ are true.

Models here are mathematical entities – they are sets of a certain kind – so a fictionalist cannot literally believe talk of logical implication or logical consistency if this is what it means. (The fictionalist can also not literally believe the talk about a set of sentences Γ, but this is easier to eliminate.)

But I think that there are reasons why even a platonist should question that Tarski's definitions give anything like an adequate account of the meaning of 'logically implies' or 'logically consistent'.

One difficulty is that the proposed definition of consistency looks too strong, and the proposed definition of consequence too weak. This comes out clearly when one takes the sentences in Γ to be about sets. Suppose for instance that Γ is the set of all truths of set theory. Since all members of Γ are true, Γ should surely be consistent. But is it obvious that there should be a model in which all members of Γ come out true? Well, if there were a model whose domain was the set of all sets, and in which 'ϵ' stood for the membership relation, then the answer would surely be 'yes': since all members of Γ are true, they would be true in this model. But everyone knows that there is no set of all sets, so there can be no model of the sort just contemplated. So if the set of all truths of set theory is Tarski-consistent, it is so by virtue of some model that does not have the full set-theoretic reality in its domain (and in which 'ϵ' may not even stand for the membership relation). Why on earth should anyone believe that there is such a model?[19]

Of course, if the language in which the members of Γ are formulated is a first order language, there is a complicated argument for the existence of such a model. First, we argue that since all members of Γ are true, there can be no derivation of a contradiction from Γ; this seems *prima facie* plausible, and I will not raise any questions about it here. But second, we must go from this conclusion to the existence of a model that makes all members of Γ true. And at this stage, the arguments (by Gödel, Henkin, etc.) are quite complicated (they are variations on proofs of the Skolem–Lowenheim theorem), and the models of set theory they produce are quite unnatural (for instance, in being countable, and in there being no guarantee that what gets assigned to 'ϵ' looks very much like membership). The fact is that it is only by virtue of an 'accident of first order logic' that the Tarskian account of consequence gives the

[19] It is no good objecting that one can allow models to be proper classes instead of insisting that they be sets. Yes, one can do that in a set theory that recognizes proper classes, like Gödel-Bernays; but then there will be no class of all classes, in which case it is unobvious why there should be a proper class model for the set of all truths about *classes*.

intuitively desirable results. The same, of course, holds for his definition of implication. If S is a sentence that is false in set theory, then set theory shouldn't imply S. But if no set – no proper part of the set-theoretical reality – could serve as a model for the entire set-theoretical reality, as seems *prima facie* possible, then set theory would Tarski-imply S. Again, if we are platonists we are saved from this possible extensional divergence between implication and Tarski-implication, by virtue of a close relative of the Skolem–Lowenheim theorem (viz., the completeness theorem). But if we demand much more than extensional adequacy in a definition, then even if we are platonists the Tarski definitions look inadequate. (And if we want to contemplate logics other than first order logic, we should be prepared for the possibility that the Tarski-definitions do not even extensionally capture the intuitive notions of implication and consistency.)

How then should we understand implication and consistency? One possibility is to take them as primitive (or at least, to take one of them as primitive: the other is definable in terms of it). Kreisel (1967) has pointed out that on this basis one gets a much more satisfying understanding of the significance of the completeness theorem for first order logic than one gets by taking the notions of implication and consistency to be defined in the Tarskian manner. The idea is that it is intuitively a *necessary* condition on Γ implying B that there be no model in which all members of Γ are true but B is not. Moreover, any intuitively sound formalization of first order logic supplies us with a *sufficient* condition of Γ implying B: in the case of Hilbert-style formalizations, the sufficient condition is that B be derivable from Γ in that formalization. So if Γ derives B, then Γ implies B; and if Γ implies B, then Γ Tarski-implies B. What the completeness theorem for a given formalization does is prove (platonistically) that if Γ Tarski-implies B then Γ derives B; this together with the previous chain shows that all three notions coincide extensionally for first order logic.

Note that on this analysis, implication is neither a proof-theoretic notion nor a semantic notion. It is a primitive notion, just as negation and conjunction and universal quantification are primitive notions. To say that those notions are primitive is to say that their meaning is not to be conveyed by definition. The meaning is to be conveyed, I think, by specifying the procedural rules involved in inferring with it. In the case of implication, these procedural rules include both 'positive' rules for recognizing implications and 'negative' rules for recognizing failures of implication. The positive rules will be sufficient for showing (fairly easily) that if Γ derives B in a typical formalization of first order logic, then Γ implies B; the negative rules will be sufficient for showing (fairly easily) that if Γ does not Tarski-imply B then it doesn't imply B. These procedural rules give the meaning of implication, quite independently of any biconditional that we get from the completeness theorem.[20]

That this is the natural way to think of logic becomes still more evident when one reflects on systems of natural deduction. For systems of natural deduction are systems in which a notion of implication is primitive: usually there will be a single axiom schema, 'A→A' ('A implies A'), and a bunch of rules of inference enabling us to derive statements about implication from other statements about implication. The notion of implication appearing in these rules clearly cannot be understood proof-theoretically in terms of the derivation procedure, since it appears within the derivation procedure; and it is wholly unnatural to construe it semantically, for a natural deduction system is a proof procedure rather than a semantics.[21] (For these systems the Kreisel conception of completeness proofs carries over, of course: the sufficient condition for $A_1, \ldots, A_n$ to imply B is that it be derivable in the system that $A_1, \ldots, A_n$ imply B, and the necessary condition is as before. The completeness proof shows that if $A_1, \ldots, A_n$ Tarski-imply B then there is a derivation of the claim that $A_1, \ldots, A_n$ imply B; so that there is an extensional equivalence between the three claims '$A_1, \ldots, A_n$ Tarski-imply B', '$A_1, \ldots, A_n$ imply B', and 'there is a derivation of the claim that $A_1, \ldots, A_n$ imply B'.)

The idea that the notion of implication is primitive, rather than equivalent in meaning to a claim about models, seems highly plausible also on the basis of more simple-minded considerations. Suppose someone were to assert each of the following:

(a) 'Snow is white' does not logically imply 'Grass is green.'
(b) There are no mathematical entities such as sets.

Such a person would not appear to the untutored mind to be obviously contradicting him– or herself, in the way he or she would be obviously

---

[20] In my view, the rules that give the meaning are not beyond revision: for instance, we would certainly revise them if we found that the 'positive' rules for showing implication conflicted with the 'negative' rules for showing non-implication. This may well have happened before; our current procedural rules may well be the product of such revision in earlier rules. I don't take very seriously the possibility that our current rules might themselves turn out to need revision, but this does seem to me *in principle* possible.

[21] In a natural deduction system for first order logic one makes only limited use of the notion of implication: one makes positive statements of the form '$A_1, \ldots, A_n$ imply B', but one never negates them or otherwise embeds them in more complicated sentences. But if one wants, one can easily extend these systems so as to allow such embeddings, including embeddings within a further occurrence of the notion of implication. See Scott (1971) (which is for Gentzen-style systems rather than natural deduction systems; but the ideas are similar).

in contradiction if he or she asserted both 'Jones is a bachelor' and 'Jones is married'; but of course (a) and (b) would be in obvious contradiction if 'logically imply' was defined in terms of sets in the Tarskian manner.

I have argued that Tarski's 'definitions' of such notions as logical implication don't give the meaning of our ordinary notions; but someone might claim that his definitions nonetheless provide *theoretical accounts of* what implication and consistency and so forth are (much as 'electromagnetic radiation of such and such wavelength' provides a theoretical account of what light is; or much as someone who did not know about alternative reductions of number theory to set theory might claim that the relation 'is the unit set of' is a theoretical account of the successor relation among numbers). I do not find this a very attractive view: it seems to me much more natural to say that implication is something we don't need a theoretical account of, for it is a logical notion like negation or conjunction or identity or existential quantification. One reason this seems more natural is that for a theoretical account to be adequate, we must *at least* be able to convince ourselves that it gives the right extension to the concept for which it provides the account; and as we have seen, the Tarskian accounts may be extensionally inadequate for some logics (even assuming platonism).

If we view '$\rightarrow$' as a primitive, we have two choices: it could be a primitive predicate, or a primitive operator. In my view it makes little difference which we choose, except that if we view it as a predicate we have to take care to avoid semantic paradox. (See Kripke 1975, pp. 713–14, for a discussion of how this can be done.) To avoid such complexities, I prefer the operator treatment. (Given a truth predicate, of course, an implication predicate can be defined from an implication operator.) We can use the implication operator to define a 1-place operator 'LTrue' ('it is logically true that'): 'LTrue (A)' could be defined as '(A v -A) $\rightarrow$ A'. (Conversely, 'A$\rightarrow$B' is definable from 'LTrue', as 'LTrue(A $\supset$ B)', so it makes no difference which of the two we regard as the primitive.) Evidently 'LTrue' should obey the laws 'LTrue(A) $\supset$ A' and 'LTrue(A $\supset$ B) $\supset$ (LTrue(A) $\supset$ LTrue(B))'. It also seems natural to allow that 'LTrue' be attached to sentences that themselves contain 'LTrue'; if so, we will presumably want to take 'LTrue (A)' as a logical axiom whenever A is a logical axiom, and to take 'LTrue (A) $\supset$ LTrue (LTrue (A))' as a logical axiom. The laws we have just found for 'LTrue' have, of course, a recognizably modal character: they are the characteristic S4 laws for a necessity operator. (I don't say that these exhaust the relevant laws. Near the end of this section I will discuss the issue of whether the S5 axiom should also be accepted.) So the procedure that I've argued is natural, of taking implication as neither syntactic nor semantic but primitive, leads to the idea that it, and related notions like logical truth, are modal notions of a certain kind.

But the operator 'LTrue' that emerges from this viewpoint is much more innocuous than most modal operators, in a number of ways. For one thing, it is thoroughly anti-essentialist. Indeed, there is no compelling reason for taking it as meaningfully applying to formulas with free variables: we can simply disallow quantifying in. It is slightly more convenient, actually, to allow the operator to apply to such formulas, by regarding a formula with free variables as logically true if and only if its universal closure is. This leads to quantifying in, but only trivial quantifying in: '$\exists x\, \text{LTrue}\,(F(x))$' and '$\forall x\, \text{LTrue}\,(F(x))$' are each equivalent to '$\text{LTrue}\,(\forall x F(x))$'. No one could possibly object to this as essentialist!

I have been arguing that a modal view of implication, logical truth, and so forth is virtually inevitable, even for a platonist. In essay 3 below, I also argue for a modal view of these notions; but the presentation there puts more emphasis on the advantages that such a view has for the anti-platonist, and this has led some to dismiss the use of logical modality in an account of implication as an *ad hoc* trick. I hope the remarks above serve to correct this.

It is worth saying something about how the ideas developed here apply to notions of implication and logical truth for logics too powerful to have complete proof procedures: logics like second order logic, or first order logic with an additional quantifier 'there are infinitely many'. Given a reasonable proof procedure for a fragment of the logic, we still want to say that a derivation of B from A suffices to show that A implies B; and given a reasonable notion of model for such a logic, we still want to say that the existence of a model in which A is true and B isn't suffices to show that A *doesn't* imply B. But there will be no completeness theorem in the offing to enable us to show that implication is extensionally equivalent to either a proof-theoretic notion or a model-theoretic notion. Should we nonetheless hold that it is extensionally equivalent to one or the other?

I believe that the natural platonistic attitude is to refrain from identifying implication even extensionally with derivability in this case – and in fact to deny such an identification. I also think that the natural platonistic attitude is to refrain from identifying implication even extensionally with the model-theoretic notion, unless there is some special reason, like the reason provided in the case of first order logic by the completeness theorem, for thinking that in the case of set theory all failures of intuitive implication will be reflected in the restricted portion of set-theoretic reality given in a model. Nonetheless, the implication relation seems intuitively to be 'closer to' the model-theoretic notion in its extension than to the proof-theoretic. If this is right, then 'logically true' should be sharply distinguished from 'logically knowable': what is logically knowable is presumably reflected in some ideal system of proof (or anyway doesn't go too far beyond this), so the attitude just suggested involves the idea that not all logical truths are knowable. Let's call this conception of logical truth (and logical implication) the broad conception.

There is, though, another possible platonist attitude to logical truth, and that is to identify the logically true with the logically knowable. This would not involve identifying logical truth extensionally with derivability in a given system, for one might not be certain that any specific system included everything that was logically knowable; but it would mean that one would probably not deny the extensional identification of logical truth with derivability in a given system unless one accepted a more powerful system, and that one would (in the case of powerful logics like those under discussion) deny the extensional equivalence of logical truth with a model-theoretic notion (even independently of considerations about the results of applying the model-theoretic notion to set theory). Let's call this the narrow conception of logical truth (and logical implication).

The dispute between the broad and narrow conceptions of logical truth and logical implication needn't be thought a substantive one: a platonist could recognize both notions as important. (The narrow notion of logical truth, which identifies logical truth with logical knowability, seems more aptly called 'absolute provability'. Reinhardt 1985 studies a somewhat similar notion under that name: but he is concerned with a notion of provability in arithmetic, rather than provability in higher order logic or some other non-axiomatizable logic.) Perhaps there are also interesting intermediate notions. Which notion one has will affect which logical laws one adopts. For instance, suppose that 'A' is a sentence in the powerful logic in question whose truth clearly doesn't turn on physical facts. (In the case of second order logic, any sentence that contains no individual constants or predicate constants, and that has no implications about how many individuals there are, fits this description.) Then an advocate of the broad conception of logical truth will be willing to accept '(A $\supset$ LTrue (A)) & (-A $\supset$ LTrue (-A))', even if he or she has no idea which of 'A' or '-A' to accept and has no reason to think that anyone could ever have reason to decide between them. If 'LTrue' is taken as meaning 'absolutely provable', though, such a claim could not be accepted.

Note that to accept the broad sense of 'LTrue' as intelligible is not necessarily to believe that there is a definite fact of the matter as to whether it is A or -A that is both true and LTrue in such a case. I think that a reasonable attitude to take is that logical notions are made clear by the rules that govern our use of such notions; and that this set of rules will always be recursive. Since a second order quantifier has no recursive proof procedure, this viewpoint has as a consequence that there is a certain vagueness or indeterminacy in second order quantification. But even so, one can consistently hold that '(A ⊃ LTrue (A)) & (-A ⊃ LTrue (-A))' always holds for such pure second order sentences: one simply says that 'LTrue(A)' has a vagueness or indeterminacy correlative to that of A.

The issues I have been discussing have a bearing on the logic of '→' and 'LTrue' even when the base language to which they are applied is purely first order. The reason is that even for first order A, there is no proof procedure adequate to sentences of form LConsistent (A) ('it is logically consistent that A'), where 'LConsistent (A)' means 'not LTrue (not A)'. If we adopt the broad conception of logical truth, we will nonetheless assert all sentences of form 'If LConsistent (A) then it is LTrue that LConsistent (A)' (by the first conjunct of the schema given in the previous paragraph). This is the characteristic axiom of S5. Accepting this axiom is consistent with saying that the notions of L-consistency and L-truth themselves have a certain vagueness or indeterminacy, by the argument of the previous paragraph: one is simply saying that however vague or indeterminate they may be, the S5 axiom is one of the principles that govern them. On the other hand, if one prefers to take 'LTrue' in the narrower sense of 'absolute provability' one will reject the S5 axiom. (And if one regards both notions as intelligible, one may want to introduce two modal operators, an S5 operator and an S4 operator, and to study their interactions: in particular, study which sentences involving the S5 operator are logically knowable.)

In the essays below I have implicitly opted for the broader notion. This is most obvious in essay 3, where I explicitly assume that the operator 'it is logically true that' obeys the laws of S5. In fact, the only real use of the S5 axiom in that paper is in connection with the justification of the 'modal soundness principle';[22] but there the use of the broader conception is probably essential, for it is not at all clear that the modal soundness principle should hold on the narrower conception of logical truth, and if not, some important claims in the paper would need modification were one to want to use only the narrower conception. Also, in several other papers I remark that the important notion of conservativeness is to be construed modally, and I contemplate the possibility of applying this in the context of non-

---

[22] In that essay I also accept certain claims of form '◇A' as axioms, and pass to '□◇A'. But I take it that this is not the use of the S5 axiom, but the principle that if B is an axiom then so is '□B'.

axiomatizable logics. It is not hard to see that the discussion of
the distinction between 'syntactic conservativeness' and 'semantic
conservativeness' in essay 4 applies to the distinction between modal
conservativeness on the narrow conception and modal conservativeness
on the broad conception: in other words, the modal conservativeness
claim cannot be accepted in the context of non-axiomatizable logics, if
the modality is read narrowly. I do not myself take this as an argument
against the use of non-axiomatizable logics, since I see nothing
unintelligible about the broad conception of logical truth; but I will not
insist on this since, as noted in essay 4, the use of non-axiomatizable
logics in physical science is now something I would rather avoid.

6 Nonlogical Modalities

I have defended the need and intelligibility of what are in effect modal
operators of a certain sort: a 'necessity' operator 'it is logically true
that', and other operators definable from it alone, such as a logical
implication operator and a 'possibility' operator 'it is logically consistent
that'. As I have mentioned, these operators are thoroughly non-
essentialist, and in this regard are quite atypical of the modal operators
typically discussed by philosophers.

A recurrent theme in the last two essays in this volume is that less
austere notions of possibility and necessity need to be treated with
extreme caution. I do not say that no sense can be made of less austere
modal notions: on the contrary, it is often possible to give fairly
'hygienic' explanations of modal notions (even essentialist ones) by
explaining them in terms of logical possibility and other fairly clear
notions. For instance, one might explain 'It is physically possible that
A' as 'A is logically consistent with all true physical laws' – I forgo a
discussion about some of the complications in the formalization of this.
I do not contend that such a definition makes the notion of physical
possibility entirely clear, for the notion of physical law is not entirely
clear. (Is the non-existence of tachyons a law, assuming it to be true?)
But for many purposes it makes it clear enough.

Unfortunately these hygienic explications of modal notions often do
not serve the purposes that friends of modality are inclined to put them
to. For instance, one use to which the less austere modalities are
frequently put is to serve as surrogates for ontology: one hears that
physical space should be dispensed with in favour of geometric
possibility, or mathematical objects in terms of mathematical possibility,
or whatever. The first question that should arise about such proposals
is about the concept of possibility being employed. Physical possibility
is consistency with *true* physical law. Are geometric possibility and

mathematical possibility consistency with *true* claims about physical space and with *true* claims about mathematical objects? If so, it is hard to see how we avoid an ontology of physical space, or of mathematical objects: for the objects seem to be needed in the account of this sort of truth. And if not, it is important to specify exactly how the notions of possibility *are* being understood; without this, one cannot evaluate the proposals for dispensing with the ontology in question.

It is not just in connection with ontology that a clarification of the less austere modalities seems important: I think that the vast majority of philosophical disputes about modality are largely verbal, turning on different uses of the modal operators by the protagonists in the dispute. More importantly, I think that nearly all attempts to apply modality to get non-modal conclusions turn on equivocation between different senses of an unclarified modal notion. To say that the modal notion one has in mind is *metaphysical* possibility is to give a pseudo-clarification, the practical effect of which I think is to serve as a licence to equivocate.

As an illustration of this cynical diagnosis of talk of metaphysical possibility consider Leibniz's modal argument against substantivalist views of space-time. (Here I will be elaborating a footnote in essay 6 that many readers have found cryptic.) Substantivalism, I take it, is the claim that space-time regions exist as entities in their own right, not as logical constructions out of matter. Leibniz thought that we could use modal considerations to argue against this ontological thesis. He asks us if we can make sense of the idea of a possible world distinct from our own, but just like ours except that throughout history everything has been shifted one mile away in some specific direction. His answer is 'no': any possible world qualitatively like ours throughout its history would just *be* our world. But, he says, the substantivalist can't say this: the substantivalist has to regard the two possible worlds as genuinely distinct, and this seems absurd.

The argument may at first seem persuasive; but let us ask, following Paul Horwich 1978, whether the argument doesn't seem equally forceful against entities that we have no doubts about – electrons, say. Just as it seems odd to think

> (DS) that there is a possible world distinct from ours but qualitatively identical to it, differing only in that throughout history everything is shifted over one mile,

doesn't it seem equally odd to think

> (DE) that there is a possible world distinct from ours but qualitatively identical to it, differing only in that Electron A and Electron B have been switched throughout their entire history?

It seems to me that it does. But if the reality of space-time regions implies (DS), doesn't the reality of electrons equally imply (DE)?

It is hard to believe that the 'Leibnizian argument against electrons' is any good. If it is not good, is that because the existence of electrons fails to imply (DE), or is it because (DE) isn't such a bad conclusion after all? My answer is that you can say what you like: these are just two alternative conventions for talking about possible worlds. (Or about possibility: (DE) can be reformulated so as not to mention possible worlds, by saying that it would have been possible for the world to have been just like it actually is except with Electrons A and B switched. And of course there is an analogous reformulation of (DS).)

If I had to choose between these conventions, my choice would be the one that says that (DE) simply doesn't follow from the existence of Electrons A and B. It seems to me that we do not normally differentiate between isomorphic 'possible worlds' – worlds isomorphic over their entire history, not just at a particular time. Rather, we accept the Principle of Identity of Indiscernibles, *as applied to possible worlds*, simply as a matter of convention. So (DE) is false by convention; but obviously this convention does not rule out the existence of electrons. The point can be put as a point about our principles for individuating objects across possible worlds: normally I think we regard individuation as sufficiently tied to qualitative characteristics (including relational qualitative characteristics, like spatial relations to other qualitatively described objects) that if there is a unique isomorphism between possible worlds (over their entire history, not just at a moment) then we regard it as making no sense to suppose that identity goes via anything other than this isomorphism. (If there are multiple isomorphisms between one world and another (due to complete symmetries in the worlds), I think that we normally think that there is no fact of the matter as to which isomorphism is the transworld identity; but we regard it as making no sense to suppose that the identity goes by anything other than one of the isomorphisms.)

It might be thought that Kripke 1972 casts doubt on the idea that this 'qualitative' viewpoint is our normal convention for transworld identity, but I do not think that is so: what Kripke's examples show is that our normal transworld identifications don't go by considering only qualitative similarity *at a time*. That is, suppose there is a possible world which is exactly like ours up until after the birth of Nixon, but which then diverges: in it, the person X born of the people qualitatively like Nixon's parents ends up looking different and having a different career and personality than Nixon does in the real world; and someone else Y ends up with the looks, character and career of Nixon. Kripke's point is that we individuate things across worlds in such a way that the isomorphism of the initial segments of the worlds (up until after Nixon's birth) counts as identity; this settles that it is X, not Y who is Nixon, so the *local* qualitative similarity of later stages of Y to Nixon is overruled. The idea that we normally regard transworld identification as independent of qualitative characteristics (even relational ones) has no support from Kripke's arguments. (I don't know if it was intended to.) Even Kripke would regard it nonsensical to suppose there a world isomorphic to ours in respect of all qualitative characteristics, even relational ones, but in which the (presumably unique) isomorphism maps Nixon into someone else.

Nor does it help to say that we don't look at possible worlds through telescopes and make transworld identifications on the basis of what we see, but that instead we stipulate worlds. For my point can be recast into the language of stipulation: as we usually talk about possibility, considerations of qualitative similarity provide limits on the possibilities we can stipulate. We can't stipulate a possible world completely isomorphic to ours in which Nixon there is like Humphrey here throughout his history (even in having parents, grandparents, etc. that are just like Humphrey's parents, grandparents, etc.). We can't stipulate a world completely isomorphic to ours in which Electron A is where Electron B in the real world is throughout its history, and vice versa. (Here I'm assuming not only that B is distinct from A in the real world, but also that no complete symmetry of the real world with respect to all qualitative characteristics maps A on to B.) And we can't stipulate that there is a world completely isomorphic to ours in which space-time region A has the properties that space-time region B has here; for instance, in which A has the properties (containing a 450 pound man, say) that the region one mile away from it does here. (Again, I'm assuming that there is no complete symmetry of the real world with respect to all qualitative characteristics – or at least, none that takes A on to B.)

It is evident that if we employ this qualitative criterion of transworld identity, or this qualitative constraint on what worlds can be stipulated, then (DS) is as false as (DE), and no more follows from the existence of space-time regions than (DE) follows from the existence of electrons.

But there is no need to insist on this qualitative standard of crossworld identification: in my view, possible worlds are just fictions, and one can speak of them as one likes. (Perhaps it would be more to the point to say that there are multiple conceptions of possibility and of possible worlds, and one can employ whichever one likes.) Thus one can perfectly well say that (DE) is true: the isomorphic worlds in the electron case are genuinely distinct. Indeed, this is the position that one would take if one were to take 'possible' in (DE) to mean *logically* possible, i.e.

formally consistent. It is also the position that Horwich takes. He argues that if it sounds counterintuitive to assert (DE), that is because usually when we have multiple possible worlds there is an epistemological problem as to which one we're in. But, he points out, in this case an epistemological problem can't arise: if there were an epistemological problem it would have to be 'How do I know that Electron A isn't where Electron B actually is instead of where Electron A actually is, and vice versa' – hardly worrying (since settlable by appeal to the meaning of 'actually'). Now if this solution is acceptable in the electron case, it is equally acceptable in the space-time case: we can say that (DS) is false, but that its falsity is not worrying because the only 'epistemological problem' that it licenses is, 'How do I know that everything isn't one mile away from where it actually is?'. To say that (DS) is true strikes me as a slightly unnatural way of talking, but I think it just a matter of convention whether one talks this way or talks in the way suggested in previous paragraphs. What I do want to insist is that whichever way one talks in the electron case, one can just as well talk that way in the space-time case, and for the same reasons.

One often hears the view that the Leibniz argument undercuts the idea of space-time regions as 'individuals' or 'substances', but allows the view that they are properties of objects. This seems to me roughly backwards. I think that the Leibniz argument may actually have some force against the view that regions are properties of matter – see the end of footnote 15 of essay 6. But, I have been arguing, it no more shows that space-time regions are not substances than it shows that electrons are not substances. Any appearance to the contrary rests, I think, on a vacillation between a sense of 'possible' in which (DS) follows from the existence of regions and a distinct sense of 'possible' in which (DS) is absurd. That these senses of 'possible' must be distinct can be seen from the electron example. When we employ the Leibniz argument to something that we may have less confidence about, like space, it is easy to confuse the senses of possibility so that we end up thinking the argument sound.

I have had an ulterior motive in defending substantivalism against Leibniz's argument: a substantivalist viewpoint is important to my position on the applications of mathematics in physical theories. But my main point has been to illustrate my contention that the use of modality to draw non-modal conclusions almost always turns on an equivocation in the modal concepts employed. Various other illustrations could be given of the same point. (Descartes' argument that he was not a material being is an obvious one, as are the modal versions of the ontological argument for the existence of God.) Some illustrations that are specially relevant to the topics of this volume can be found in the last two sections of essay 6, in section 2 of essay 7, and in section 7 of the same essay.

There is one final instance that I would like to comment on. In Hale (1987) and (drawing heavily on Hale) Wright (forthcoming), it is alleged that modal considerations undermine my version of anti-platonism. Hale and Wright both note that I regard mathematics, and the existence of mathematical entities, as consistent (indeed, as having a strong form of consistency that I call conservativeness); and that I take consistency as a primitive modal notion, a sort of possibility. They then argue that since I regard it as false that there are mathematical entities, I must hold the existence of such entities to be 'contingently false'; and they both proceed to interpret 'contingently' in some non-logical sense (i.e. some sense other than 'neither logically true nor logically contradictory'), to make the position seem absurd. For instance, Wright says

Field has no prospect of an account of what the alleged contingency is contingent *on*. This world does not, in Field's view, but might have contained numbers. But there is no explanation of *why* it contains no numbers; and if it had contained numbers, there would have been no explanation of that either. There are no conditions favorable for the emergence of numbers, and no conditions which prevent their emergence. (pp. 46–7)

(Wright describes this as 'the Achilles' heel of Field's position' (p. 26).) It should be noted that if this objection were good, it would apply equally well against any platonist who was not a logicist: that is, any platonist who agreed that mathematics goes beyond mere logic, and hence that the denial of mathematics is logically consistent and hence 'contingent'. But of course it is not good, for as I've said it turns on an equivocation on the meaning of 'possible'. If anyone doubts this, I suggest that they try out the analogous argument in theology.[23]

To be fair, I should note that in both Hale's and Wright's presentations, the above argument is made to look a bit better by being intertwined with two other more interesting arguments.

Their first supplementary argument is that without the assumption that mathematics consists of necessary truths, the view that mathematics is conservative looks unjustifiable. (A related argument would be that

---

[23] 'Surely the existence of God is logically consistent, so if there is no God, it is *contingently* false that there is a God. But the atheist has no prospect of an account of what this alleged contingency is contingent *on*. There is no explanation of *why* the world contains no God, and if it had contained one, there would have been no explanation of that either. There are no conditions favourable for an emergence of God, and no conditions that prevent His emergence.' This is in effect what is known as Anselm's second ontological argument.

without the assumption that mathematics is true, the view that it is consistent looks unjustified.) The idea behind this argument is that a platonist who holds that mathematics is in some sense necessary *can* justify the view that it is conservative: the justification is simply that conservativeness follows from necessary truth. (Similarly, the platonist can justify his or her belief that mathematics is consistent, by noting that consistency follows from truth.) The obvious reply, of course, is that one can in a similar sense 'justify' any belief by providing a logically stronger belief from which the first follows. What would be necessary for Wright or Hale to sustain their point would be to show that the platonist has better reasons for the view that mathematics is necessary (or true) than the anti-platonist has for the view that mathematics is conservative (or consistent). It is not *out of the question* that this could be defended: sometimes the best way to argue for a logically weaker position is to argue for a logically stronger position; sometimes the logically weaker position when separated off from the logically stronger one looks *ad hoc*. (An example, I would argue, is the view that the observational consequences of a theory are true, separated off from the view that the non-observational consequences are true.) But neither Hale nor Wright offers the slightest reason for thinking that this is so in this case, and I have been unable to construct a plausible argument for the claim.

Their second supplementary argument is that anyone who holds both that the existence of mathematical entities is 'contingently false' and that mathematics is conservative can offer no reason *not* to believe in mathematical entities. More fully: the conservativeness of mathematics means that any internally consistent combination of nominalistic statements is also consistent with mathematics. Consequently, the argument goes, no combination of nominalistic statements can provide reason for or against the belief in mathematics. So how can there be any reason not to believe in mathematics? 'It seems, then, that Field has no choice but to admit that he has *no* evidence for his nominalism . . . It follows that he ought not to be a nominalist but an agnostic.' (Wright, p. 44.) The reply to this, of course, is that Wright is ignoring the relevance that I claim for issues of dispensability and indispensability. The conservativeness of mathematics does not in itself show that there can be no reason to believe mathematics: as I have repeatedly stressed from the time of my book *Science without Numbers*, to combat the argument for platonism one must also show mathematics dispensable (in science, and as I have more recently emphasized, in areas such as metalogic as well). The other side of the coin, it seems to me, is that if the dispensability programme *can* be carried out, that gives us reason to not literally believe mathematics but only to adopt a fictionalist attitude towards it. Admittedly, we can't have *direct evidence* against mathematical entities. We also can't have direct evidence against the hypothesis that there are little green people living inside electrons and that are in principle undiscoverable by human beings; but it seems to me undue epistemological caution to maintain agnosticism rather than flat out disbelief about such an idle hypothesis. I think that platonism has seemed a plausible position because it has been assumed that the existence of mathematical entities is *not* an idle hypothesis. But if it can be shown that the hypothesis is dispensable without loss (in explanations, in descriptions of our observations, in accounts of metalogic, and so on), then I think it natural to go beyond agnosticism and assert that mathematical entities do not exist. I suppose there is no need to insist on this: perhaps agnosticism would be a sufficient conclusion. In any case, there is nothing in this supplementary argument, nor in the previous one, that can serve to rescue the modal argument on which Wright and Hale primarily depend.

7 The Anti-platonist's Resources

One frequent line of criticism of my version of the anti-platonist programme – a criticism developed at length in Resnik 1985a and 1985b – is that it is not 'genuinely nominalistic' in that I allow the use of apparatus that a 'real nominalist' cannot allow. In particular, I allow the use of an ontology of space-time, conceived as irreducible to the matter that occupies it; and I allow the use of certain logical devices that go beyond first order logic. Now, it seems to me that the issue of what is and what is not 'nominalistic' is totally without interest. I have no attachment to the label. The real issue is: is there good reason to prefer a metaphysics that does without mathematical entities to one that includes them, if in the former one has to use a space-time ontology and an expansion of first order logic? I think that there is: *especially* if (as I am inclined to argue) the space-time ontology and the expanded logic are things that *even the platonist* needs to employ.

I do not intend to discuss here the complicated set of issues about space-time and about logic required for a decision on these matters – there is a good bit on this elsewhere in this volume, and in *Science without Numbers* – but a few remarks in clarification of my views may be helpful.

I'll begin with three remarks about my employment of a space-time ontology.

1. I have used a space-time ontology (Field 1980) in showing how to give mathematical entity free formulations *of physical theories*: theories

of gravitation, electromagnetism, and so on. The centrality of space-time and its structure to physical theories like these is hardly a novel feature of my nominalistic programme: on the contrary, it is part of *any* treatment of these theories, a fact that has been stressed again and again in recent work on the foundations of physical theory. (Most of the recent work that stresses this is highly platonistic.) I should stress that it is *only* in the context of such physical theories that I think it appropriate to invoke a space-time ontology. I made a point in essay 2 (p. 64) of *not* invoking space-time structure in typical applications of number theory where talk of space-time is foreign. Nor did I invoke it in essay 3, in discussion of proof theory: one could perhaps imagine a 'nominalist' saying that there is no obstacle to accepting standard proof theory as a body of truths, because unwritten derivations can be construed as derivation-shaped regions of space; but this is not a position that I regard as the least bit attractive, and it is not the position that I took in that essay.

Not only would I not be willing to employ a space-time ontology in connection with typical applications of number theory or proof theory, I also wouldn't be willing to employ it in connection with all applications of real analysis. The *most central* applications of real analysis have been in connection with empirical theories in which space-time plays a central role: for instance, theories of gravitation, of electromagnetism, and so forth. But real analysis is sometimes employed in other areas: for instance, in (quasi-)psychological theories of degrees of belief. In my view, one should only invoke space-time in connection with those applications of mathematics in which space-time is relevant: in an account of the application of real analysis to the theory of degrees of belief it would be quite out of place. (That's one reason why I do not propose that we simply translate real analysis into claims about space-time. There are other reasons too: for instance, an empirical discovery that space-time is quantized should not be construed as an empirical discovery that the real numbers are not densely ordered.) I think in fact that there is reason to doubt that the real numbers, invented to deal with physical space and time, should turn out to correctly describe the structure of a totally separate phenomenon like degrees of belief; but if that were to turn out the case, my principle that underlying every good explanation there is an intrinsic explanation would dictate that we find a part of psychological reality to use in our account of the application of the real numbers to psychology, rather than using physical space-time.

2 I would also like to reiterate that I think that most philosophers who are opposed to employing a space-time ontology in giving an anti-

platonist treatment of physics rest their opposition on an outdated conception of physics. I have noted in several essays in this volume that space-time regions are now known to be genuine causal agents: that is what field theories like classical electromagnetism or general relativity or presumably quantum field theory tell us. When Resnik (1985a) writes

the physicist's question has never been 'What properties of space-time points are responsible for these phenomena?' but rather 'What is the structure of space-time?' (p. 167),

I think he is wrong: a theory of the electromagnetic field is precisely a theory of the properties of even those parts of space-time unoccupied by physical objects. You can verbally deny this by viewing fields as entities rather than as modes of space-time; but then what I mean by a region of space-time is just what you would mean by a part of a field; so that in your terms, it is only the parts of the field that are needed in nominalistic physics, and hence it is the legitimacy of appealing to parts of the field that the critic of nominalism must deny. (Resnik's remarks seem to indicate that he thinks of fields as a manner of speaking, to be replaced by talk of interactions between matter alone: see for instance pp. 166–7 of his 1985. But the feasibility of this 'action-at-a-distance' viewpoint is doubtful. Resnik also seems (on pp. 167–8) to advocate some sort of modalist view of unoccupied space-time regions, of the sort I cast doubt on in essay 6.)

3   Finally, some comments on the fact that the space-time ontology I invoke in connection with physical theories is an ontology of *regions*, and includes minimal regions of zero size, i.e. *points*. Some have objected that even if points are OK from an anti-platonist perspective, regions aren't, for they are just sets of points. Part of the purpose of the opening section of essay 6 is to cast doubt on the idea that regions should be viewed as sets of points. I view this as having no more plausibility than the view that ordinary physical objects are sets whose members are the points of matter that make them up. Points of matter are in fact somewhat more dubious entities than are ordinary objects; in exactly the same way, points of space-time are somewhat more dubious than regions of finite size. I see some virtue in trying to eliminate the appeal to points, using only bigger regions; I see no virtue in eliminating regions other than points if points are retained.

I would not be terribly surprised if it were to turn out possible to give physical theories in some sort of point-free geometry – indeed, a geometry that did not rely on points anywhere in the development. (Perhaps the space would be isomorphic to the set of regular open sets in the usual geometry; or to the quotient space got by factoring the usual geometry with the ideal of sets of measure zero; or some such thing.) It would certainly be no *easy* task to do this: apart from the difficulties in developing the geometry itself with no mention of points, there is the further task of developing analogues of the theory of scalar fields, tensor fields and so forth. But it seems to me quite conceivable that it is a performable task. If such a task is in fact possible, I think it would be of some interest: it would show that we need to attribute much less structure to space-time than we now think.

And that would be attractive, for a variety of reasons. For instance, a number of people have observed that the continuum hypothesis has an analogue for physical space, statable with purely physical resources (i.e., without set theory): it will say that for every subregion of a line, there is either a graph portraying a 1–1 correspondence to another line or a graph portraying a 1–1 correspondence to a discrete region of type ω. The difficulty in deciding the continuum hypothesis presumably applies to its physical analogue as well. (And there is no obvious reason why the answer to the mathematical question, if there is a unique answer, should have to be the same as the answer to the physical question: for there is no obvious reason to suppose that lines in space should have to be isomorphic to the real numbers or that regions should have to be in 1–1 correspondence to sets of points.) It may seem hard to believe, though, that the question of whether there are subregions of a line with intermediate cardinality can be 'physically real'. If this does seem hard to believe, then the development of a point-free physics should be welcome, since presumably no such physical analogue of the continuum hypothesis would arise in it.

Getting rid of points would also remove any temptation (already inappropriate) to view regions as sets of points. In general, it would help reduce the (already inappropriate) tendency to view appeal to space-time as appeal to mathematical entities in disguise. At present, though, no one (nominalist or platonist, substantivalist or relationalist) knows how to dispense with points (of both space and matter simultaneously) within physical theory.

Enough on space-time. I close with four comments on my employment of logical devices that go beyond first order logic.

1   I have tried to argue that *even the platonist* needs these devices: in section 5 of this introduction and elsewhere I argue that even the platonist must understand consistency as a primitive notion (in effect, a modal notion); and in section 4 of essay 7 (and to a lesser extent, the postscript to essay 3) I argue that even the platonist needs some sort of

device of (restricted) infinitary conjunction not definable in first order logic.

2 It isn't really important what counts as logic and what doesn't. What is important is the use of various operators (perhaps a consistency operator, perhaps a device of infinitary conjunction, perhaps a quantifier 'there are only finitely many'), taken as not needing definition, but governed by a body of primitive assumptions and primitive rules of inference. By taking the operators as primitive, we avoid having to employ the set-theoretic definitions that engender ontological commitment. I think it is natural, if one has the primitive operators, to call them logical, but this lexicographical decision is not what is playing the substantive role.

3 One issue that has worried many (including myself at one time) about the use of these operators not found in first order logic is that the principles that govern them have no complete proof procedure. This seems *prima facie* worrying quite independently of anti-platonism: many people, myself included, think that the only sensible view for either a platonist or an anti-platonist to take of our understanding of basic logical operators is that it consists of our ability to reason with these operators in accordance with various procedural rules; and this seems somehow to conflict with the use of logics with no complete proof procedure.

I think it is easy to provide a satisfactory reply to this worry, but before giving it I will mention two reasons for thinking, even in absence of the reply, that the worry can't be right. In both cases, the point is that if the worry were right it would undermine too much.

Point one: I think it is clear that the procedural rules that are important to our understanding include not only 'positive' rules for ascertaining that one thing follows from another, but also 'negative' rules for ascertaining that one thing *doesn't* follow from another. This suggests that if it is worrying to employ any logic with no complete set of positive rules, it ought to be worrying to employ any logic with no complete set of negative rules too. That of course would rule out even first order logic. (A moral one might prefer to draw is that *either* a complete set of positive rules *or* a complete set of negative rules is enough for legitimacy. But it would need an argument why one was enough. Perhaps the Kreisel argument of section 5 provides one, but if so, it works by embedding the logic with a complete set of positive rules only (first order logic) in a richer logic (employing a notion of logical implication or logical consistency) which has neither a complete set of positive rules nor a complete set of negative rules.)

Point two: if the argument works to show the illegitimacy of *operators* with no complete proof procedure, why shouldn't it equally show the illegitimacy of *predicates* with no complete proof procedure? Why shouldn't it, for instance, show the illegitimacy of the predicates 'set' and '∈' (on grounds quite distinct from any that an anti-platonist would offer)?

If anyone were to seriously raise this last argument as an objection against sets, I take it that there are two possible platonist replies. Perhaps the most popular among platonists would be to insist that we have perfectly clear notions of set and membership, which outrun the procedural rules that we employ. This of course would violate what I called above 'the only sensible view of understanding'. But another alternative, that doesn't violate a procedural conception of understanding, would be to admit that the notion of set has a certain amount of vagueness or indeterminacy, but to say that that does not make it illegitimate.

The same two options are available in the case of operators. The first, of course, violates 'the only sensible view about understanding', so we are left with the second: admitting that there is a certain amount of vagueness or indeterminacy in our understanding of operators like 'there are only finitely many' or 'it is consistent that', since the procedural rules that govern them leave certain questions open; but saying that they are nonetheless clear enough to use. Why should that position be so bad? Indeed, isn't that what the sensible platonist says too? The sensible platonist proposes a set-theoretic explication of these operators within a recursively axiomatized set theory in first order logic; but the procedural rules for his or her primitives, 'set' and '∈', don't determine a unique extension for them, so a definition of the other operators in terms of these primitives won't yield a determinate extension for the operators either.

The real question, I think, is not whether it is legitimate to employ operators with no complete proof procedures, but whether sufficiently strong partial proof procedures are available to the anti-platonist to serve the purposes he or she requires. It would be a legitimate complaint about *Science without Numbers*, and about the essays here, that I do not say anything about what those partial proof procedures are (except in the case of the logical consistency operator). Indeed, we need not only a sufficiently strong partial proof procedure – a sufficiently strong set of positive procedural rules – we need a sufficiently strong set of negative rules as well; again it would be legitimate to complain of my silence on what these are (except in the case of logical consistency). A specification of the procedural rules *is* ultimately needed. Here I can only say that the presence of a primitive consistency operator (or implication operator) is likely to help with the other operators: a decent set of procedural rules governing a cardinality quantifier and a consistency operator together is likely to be easier to come by than a decent set of rules governing the former alone. Perhaps this is evident from section 5.

Some may object that we can only come to understand logical operators from either a definition in previously understood terms or a model-theoretic account of their contributions to truth conditions; and that in the case of 'there are finitely many' or 'it is consistent that', the platonist can provide a definition or a model theory in terms of sets, whereas the anti-platonist cannot provide either one. I think, though, that it is patently false that we can only come to understand logical operators in these ways: that isn't how we come to understand the negation operator or the existential quantifier, for we can't define them in more basic terms, and any model theory for them contains the very operators. The proper platonist analysis of the significance of model theory is the one sketched above in section 5 (and in essay 3 below, where it is shown to be modifiable in an anti-platonist direction): and I have already indicated in section 5 how the analysis works for non-axiomatizable logics.

4   I have adopted a somewhat experimental attitude towards which logical devices to employ. In *Science without Numbers*, I toyed with (but expressed considerable dissatisfaction about) something I called 'the complete logic of Goodmanian sums'. ('The complete logic of the "part of" relation' would have been a better name for it.) I also toyed (much less hesitantly) with employing a primitive 'there are only finitely many' quantifier, despite the fact that (as I noted in *Science without Numbers*, p. 127) such a quantifier was otiose for the physical theory under discussion there in the presence of the 'logic' of the part/whole relation. The finiteness quantifier (as well as many more complicated cardinality quantifiers) would also be otiose in the presence of many devices of (restricted) infinitary conjunction, for instance many forms of substitutional quantification. My attitude has been that we ought not to regard first order logic as having been laid down from on high as the logic within which all theorizing must take place; rather, we should experiment, seeing what it takes to produce the most satisfactory overall theory. (In essay 4 I have recanted on the 'logic' of the part-of relation, and noted that even without it we don't really need the finiteness quantifier for physics.) Many critics of my programme have portrayed me as asserting the need for *each* of the extensions of first order logic that I have discussed, but that has never been my attitude. Nor, of course, is my attitude that whenever we run into trouble with the nominalization programme, we can feel free to invent some fancy new

operator that solves the problem by fiat. My attitude, rather, is that the nominalization programme has not been pursued far enough for confidence about what the best choice of a logical basis is. Ultimately we will need to make a choice, and it should be one that employs only a few natural devices that go beyond first order logic – preferably, they should be devices for which it can plausibly be argued that the platonist needs them as well. But until we are farther along, progress will only come if we experiment.

There are, of course, no hard and fast rules that can be laid down in advance about what counts as a 'simple and natural' extension of first order logic. Those of us interested in developing the anti-platonist programme should do the best we can; once we see more fully what can be done with what resources, we will be in a position where we can make an informed decision about whether any extra resources we must employ to avoid mathematical entities are worth the price. I certainly do not think that at this point the anti-platonist view is obviously the right one: it may be in the end that our conclusion should be that the price is too high. But I do think that the essays in this volume, together with *Science without Numbers*, show the prospects for a reasonable anti-platonism to be better than they seemed to be before.

2

Realism and anti-realism about mathematics*

Realism about mathematics is the doctrine that there really are mathematical entities – numbers, functions, sets and so forth. Why should anyone believe this doctrine?

One possible answer is this: the only known account of mathematical truth not subject to obvious difficulties is the same account of truth that works outside mathematics. (Tarski's account, essentially.) According to this account of truth, the sentence 'There are prime numbers greater than seventeen' is true only if there is at least one entity with the properties of being a number, being prime and being greater than seventeen; and such an entity would obviously be a mathematical entity. I have heard it argued that unless we can come up with an alternative account of mathematical truth, this consequence of the standard account of mathematical truth forces us to believe in mathematical entities.

In fact, of course, it does no such thing. If we assume this account of mathematical truth – and I have none better to offer – then we can't *both* reject the existence of mathematical entities *and* regard standard mathematical theorems like 'There are prime numbers greater than seventeen' as true. But then the question is, what reason is there to regard the standard mathematical theorems as true? It is hard to see why the assumption that standard mathematical theorems are true should be thought to be more obvious and less in need of defence than the assumption that mathematical entities exist.

It is sometimes said that the only kind of reason there could be for not regarding standard mathematical theorems as true is a specifically mathematical reason; that is, a reason of the sort best evaluated by mathematicians, not philosophers. The idea behind this is evidently that

* This paper was prepared for the Rice University Conference on Realism and Anti-Realism, November 1981.

if standard mathematics really isn't true, we ought to revise it so as to make it true, and the result would be a new mathematics; and surely it is the mathematicians and not the philosophers who are in the best position to decide if such a proposed new mathematics really is better than the old one.

But this way of arguing rests on the implicit premise that a mathematical theory must be true to be good. Suppose for the moment that there are no mathematical entities: then no mathematical assertion that begins with an existential quantifier is true, whereas every such assertion that begins with a universal quantifier is (vacuously) true.[1] Obviously then if we were to revise our mathematics so as to retain the true assertions and not retain any of the false ones, the theory we would get would be of no mathematical interest. From our anti-realist's standpoint, then, truth is one thing and mathematical interest is something else, incompatible with truth. What mathematicians are in a better position than philosophers to evaluate is what counts as good mathematics; but it takes the philosophical assumption that what is good mathematics is also true to conclude that it is the mathematicians who are in the best position to decide which mathematical assertions are true.[2]

Our dispute then is between on the one hand a realist who believes that there are mathematical entities, and that the goal of mathematics is to give a true account of them and their interrelations, and on the other hand an anti-realist who does not believe that there are mathematical entities, and who thinks that mathematics has some goal other than truth.

What alternative goal might the anti-realist suggest? One goal that has occasionally been suggested as an alternative to truth is consistency. Of course, consistency by itself is no guarantee of mathematical interest; but for the mathematical realist truth isn't a guarantee of mathematical interest either, for it is only the sufficiently rich true theories that are

[1] For simplicity I've restricted my attention here to assertions with no individual constants, so as to avoid having to take a stand on the truth values of assertions containing 'empty names'. But whatever the details of one's views of empty names, it will be a consequence of the anti-realist's position that whether a mathematical assertion is true or false will depend only on uninteresting facts about its syntactic form; so the point remains that the theory containing precisely the true assertions will have no mathematical interest.

[2] It goes without saying that '2 + 2 = 4' is *true according to arithmetic*; the philosophical assumption at issue isn't that it is true according to arithmetic, but that it and the rest of arithmetic are *true*. If the philosophical assumption is incorrect, then '2 + 2 = 4' is a lot like 'Santa Claus lives at the North Pole', which is *true according to a certain myth* but is nonetheless false (as Robert Peary's observations showed). For good discussions of the analogy of mathematics to myth and fiction, see ch. 2 of Charles Chihara's *Ontology and the Vicious Circle Principle*, and Steve Wagner, 'Arithmetical Fiction'.

mathematically interesting. (The theory that there are at least two mathematical objects is, according to the mathematical realist, true, but it is certainly of no mathematical interest.) One version of anti-realism, then, is that if a mathematical theory is sufficiently rich to be interesting, the only further thing we ought to ask of it is that it be consistent; it is quite irrelevant to require that it also be true, and, in fact, this latter goal is never fulfilled since it would require the existence of mathematical entities.

I think that if one looks only at considerations internal to mathematics, it will be hard to argue against this form of anti-realism. But if we remember that one important feature of mathematics is its application to the physical world, we can see that the consistency of mathematics is definitely not enough. For a theory could be consistent and yet imply false conclusions about the physical world; obviously if a mathematical theory were to have this characteristic it would be deficient despite its consistency.

Is the situation just envisioned a real possibility? Before answering this, I should explain that when I speak of a mathematical theory implying false conclusions about the physical world, I am taking the phrase 'about the physical world' in a very narrow sense: I am not regarding an assertion as 'about the physical world' if it quantifies over mathematical entities. Thus, though the assertion that the set of all apples in this room is in 1–1 correspondence with the set $\{0,1\}$ might in some broad sense be regarded as an assertion about the physical world, it is not to be so regarded for purposes at hand. By contrast, the assertion 'there are distinct x and y that are apples in this room and are such that every apple in this room is either x or y' is purely about the physical world even in the narrow sense. To guard against the danger of equivocating between my narrow sense of 'about the physical world' and any broader sense that might be attached to this phrase, I will sometimes use the phrase 'nominalistic assertion' instead of 'assertion about the physical world'. Then my claim at the end of the last paragraph was that any mathematical theory that implied false nominalistic assertions would be inadequate, even if that theory was consistent.

Is there any chance that a consistent mathematical theory could imply false nominalistic assertions? It might seem that the answer is 'no': After all, a mathematical theory speaks only of mathematical entities, and a nominalistic theory speaks only of non-mathematical entities, so how could a mathematical theory have any nominalistic implications at all beyond logical truths?

Unfortunately, this argument is based on a false premise: that mathematical theories speak only of mathematical entities. Consider the kind of set theory we employ in practice – not pure set theory, but a set theory that allows for sets with non-sets as members. Such a set theory does not in any obvious sense 'speak only of mathematical entities': for instance, the usual such set theories will assert of each non-mathematical entity that it is a member of various sets. Admittedly, that seems like a rather weak sort of assertion to make about non-mathematical entities, but it is enough of an assertion to cast doubt on the argument in the last paragraph that because a mathematical theory 'speaks only of mathematical objects' it can't have any implications about non-mathematical objects alone beyond logical truths. (Indeed, if one tries to formalize the argument of the last paragraph by means of Robinson's Consistency Theorem or Craig's Interpolation Lemma, one sees that the fact that set theory says of each non-set that it is a member of sets is indeed enough to block the argument.)[3] So maybe a consistent mathematical theory really could have false nominalistic consequences.

And in fact it could. Consider the following mathematical theory M: start with Zermelo-Fraenkel set theory, modified so as to allow for sets with non-sets as members and so as to include the assertion that there is a set of all non-sets; and interpret the separation and replacement schemata in such a way that the empirical vocabulary that is ordinarily used to describe non-sets is allowed to appear in the instances of the schematas. (As observed in the preceding note, this broad interpretation of the schematas is the one we use in intuitive set theory: we think, for instance, that it is unproblematic to assume that there is a set of points of space at which the temperature is at least 0°C, but this could not be justified unless we allowed the term 'temperature' to appear in the schema of separation.) So far, we have intuitive set theory. Now, to obtain M, drop the axiom of infinity and replace it by its negation. This theory M is clearly consistent (indeed, it has an obvious relative consistency proof); but it has consequences about the physical world that conflict with most current physical theories.

To see this, let T be any nominalistic theory that implies the existence of an infinite discrete linear ordering of non-mathematical entities: more precisely, suppose that there is some formula $A(x,y)$ in the vocabulary of T such that from T we can derive the assertions

(i) $\forall x \forall y \forall z\ [A(x,y)\ \&\ A(y,z) \supset A(x,z)]$
(ii) $\forall x\ -A(x,x)$
(iii) $\forall x \forall y\ [A(x,y)\ \text{v}\ x=y\ \text{v}\ A(y,x)]$

---

[3]   Actually, the set theory we employ in practice differs from pure set theory not only in saying of each non-set that it is a member of various sets, but also in using empirical vocabulary in the construction of sets by the separation and replacement schemata. (E.g., we can introduce the set of all points in space whose temperature is at least 0°C.) This fact creates still deeper problems for the argument in the preceding paragraph. For details see my book *Science without Numbers*, pp. 17–18.

and

(iv) $\forall x \exists y$ [A(x,y) & $-\exists z$ (A(x,z) & A(z,y)))],

where the quantifiers are restricted to certain non-mathematical entities (not necessarily all of them).[4] It isn't hard to see that T and M together are inconsistent: for in T and M together you can define a function symbol mapping some x in the ordering into 0, its immediate successor into 1, the successor of that into 2 and so on; the fact that there is a set of all non-sets together with the replacement schema then gives you the existence of a set containing 0,1,2 and so on, in violation of the negation of the axiom of infinity. So M is inconsistent with any T of the form described, which means that M implies the denial of any such T (i.e., the denial of some finite conjunction of axioms of T). It seems quite likely, however, that some such T is true: for instance, any theory according to which physical space contains at least one infinite line, and which has the rather minimal resources required to select an infinite sequence of equally spaced points along such a line, will be declared false by the mathematical theory M.[5] I think this is enough to show that there is something terribly wrong with M, even though it is consistent.

It isn't really essential to my argument that some theory T of the sort just described is true. For it doesn't really seem essential to the criticism of a mathematical theory that it leads to conclusions about the physical world that are actually false. It seems to be an almost equal deficiency if it leads to conclusions that might well have been false: if a mathematical theory entailed that there were exactly nine planets in our solar system, all but the most unregenerate rationalist would feel that this showed that that mathematical theory was unacceptable. By this standard, the theory M above is definitely unacceptable, even if no physical theory T of the sort described above is correct. If you want to deny the axiom of infinity in set theory, you can; but in order to avoid the difficulties we've seen in M, you must make other modifications in set theory at the same time (such as imposing restrictions on the schema of replacement) so that the unwanted conclusions about the physical world will be avoided.

---

[4]   We can allow A(x,y) to contain free variables other than x and y; in this case (i)–(iv) should be prefixed with restricted or unrestricted universal quantifiers for these additional variables.

[5]   In fact, if we strengthen M by adding the axiom of choice, we can weaken the requirements on T: T then need only contain a finite number of assertions which together entail that there are infinitely many objects. For instance, any physical theory which implies with finitely many axioms that light rays have infinitely many parts would be inconsistent with the strengthened M; nevertheless, there is little doubt that the strengthened M is consistent.

We see then that a consistent (and rich) mathematical theory can be seriously deficient, so our first version of anti-realism in which consistency replaces truth as the goal must be scrapped. But now let us try a second version of anti-realism, not subject to the above objection. Let us say that for an interesting mathematical theory to be good, it need only be *consistent with every internally consistent theory about the physical world*. This is still a version of anti-realism, because the point remains that the theory need not be true to be good.

Let us call a mathematical theory that is consistent with every internally consistent theory about the physical world *conservative*. As is easily seen, this is equivalent to the following:

> (C) A mathematical theory M is conservative if and only if for any assertion A about the physical world and any body N of such assertions, A doesn't follow from N + M unless it follows from N alone.

Our modified anti-realism, then, says that besides being sufficiently comprehensive to be interesting, a mathematical theory must be conservative, but need not be true.[6]

It might initially be thought that a mathematical theory that was conservative in the sense of (C) couldn't possibly be of interest, or at least, could be of no interest in application to the physical world. For if, for any nominalistic theory N, you don't get any more conclusions from N + M than you do from N alone, what possible value could M have?

This argument, however, is faulty. Indeed, it *has* to be wrong, for as is perhaps obvious from what I said four paragraphs back,[7] there are strong reasons for supposing that standard mathematics is conservative, but no one could deny that it has value. Still, the argument challenges us to say what value a mathematical theory could have, given that it is conservative. In fact, there are two possible ways in which a conservative theory might have value.

In the first place, a conservative mathematical theory might facilitate inferences from nominalistic theories. That is, if M is conservative, N + M doesn't imply A unless N itself implies A (where N is a

---

[6] In my book I argued that the conservativeness of standard set theory 'almost follows' from the consistency of that theory; more accurately, that the consistency of that theory suffices for proving a slightly restricted version of the conservativeness of that theory. Although the gap between consistency and conservativeness is quite small for standard set theory, it is large for mathematical theories generally, as the example of the deviant set theory M makes clear.

[7] This is argued in more detail in *Science without Numbers*; see pp. 12–14 and appendix to ch. 1.

nominalistic theory and A a nominalistic assertion); but it might be much easier to see that A follows from N + M than it is to see that A follows from N alone. As we will see later, mathematics really does serve to facilitate inferences in this sense, and that is certainly a large part of its value.

In the second place, a mathematical theory that was conservative might be of value in an additional way: it might occur essentially in the premises of e.g., our physical theories. This is not ruled out by the conservativeness of mathematics; the conservativeness of mathematics says only that whenever you have a nominalistic body of assertions N, then N + mathematics doesn't have any more ultimate power as far as nominalistic consequences are concerned than N alone has; but to say that is *not* to say that it is always possible to find nominalistic bodies of assertions to do what we want. Indeed, as I will argue later, ultimately the only serious argument for the view that mathematics is true as well as conservative turns on the premise that nominalistic bodies of assertions are not always available. We see then that there are two *prima facie* possible ways that mathematics might be useful, despite its conservativeness: it might be useful in *facilitating inferences* (between nominalistic premises and nominalistic conclusions), and it might also be useful in being *theoretically indispensable* (i.e. needed in the premises of some important theory). It is crucial to the realism/anti-realism controversy whether it is useful in both ways, or only in the first.

Before getting into this question, however, I should say something about the relation between conservativeness and truth. Unlike consistency, conservativeness does *not* follow from truth; our anti-realist, then, is not really substituting a *weaker* goal in place of the realist's goal of truth, he or she is substituting a *different* goal. But the goal he or she is substituting is one that the typical realist also implicitly recognizes. For as already remarked, there are compelling reasons why the realist as well as the non-realist should regard conservativeness as a requirement of a good mathematical theory, and should believe that the mathematical theory he or she accepts meets this requirement. Indeed, I believe that the traditional realist assertion that good mathematical theories are not only true but necessarily true is more or less equivalent to the claim that good mathematical theories are not only true but also conservative. Conservativeness might loosely be thought of as 'necessary truth without the truth'. A conservative theory, like a necessarily true one, is compatible with any possible state of the physical world; the only real difference between a conservative theory and a necessarily true one is that the conservative one need not be true at all.

The fact that it is a requirement of a good mathematical theory that it be conservative marks a sharp contrast between mathematics and science. Indeed, it is a goal of a good scientific theory that it *not* be conservative. By this I don't mean merely that for a good scientific theory T there will be *nominalistic* N and A such that N won't entail A but N + T will. *That* much is rather trivial, since it is pretty clear that even if the scientific theory T doesn't itself contain non-trivial nominalistic assertions (assertions, other than logical truths, which say nothing about mathematical entities), still it will certainly have non-trivial nominalistic consequences; and letting A be such a consequence, and N consist only of logical truths, it is *trivial* that T violates conservativism in that sense.

But there is a much more interesting sense in which physical theories are not and should not be conservative. Physical theories raise issues about realism, just as mathematical theories do. In physical theories, questions of realism arise not for mathematical entities but for such entities as subatomic particles, fields, space-time regions – *unobservables*, to have a convenient word for it, though it is not clear that this word is appropriate in all cases (e.g., it is not clear that space-time regions should be regarded as unobservable). Now in the case of mathematical theories, conservatism means that assertions about the *controversial* entities don't lead to new conclusions about the *uncontroversial* entities from premises about the uncontroversial entities; and that, I suggest, is what we ought to mean by conservatism generally. Let us say then

> (C) A physical theory T is conservative if and only if for any body
>   P of assertions about observables and any assertion A about
>   observables, A doesn't follow from P + T unless it follows
>   from P alone.

Now on this more interesting understanding of what a conservative physical theory is, it remains the case that any physical theory that is conservative is totally unacceptable – unless of course it is part of a larger physical theory that is acceptable and non-conservative. Acceptability in science *ultimately* requires non-conservativeness, though acceptability may trickle down from a theory that is directly acceptable and hence non-conservative to conservative sub-theories of this theory. The reason that acceptability ultimately requires non-conservativeness, in the case of physical theories, should be obvious. For a conservative physical theory is untestable: we test a theory precisely by using that theory to derive consequences about observables (viz., experimental predictions) from premises about observables (viz., those that describe the experimental conditions). If we perform the experiment and the prediction is borne out, that to some extent confirms those premises of the theory that were essential to the deduction of the 'observational' conclusion from the 'observational' premises.[8] But if a physical theory were conservative in the sense just explained, none of it could ever be essential to the derivation of conclusions about observables from premises about observables, so none of it could ever be confirmed (except as part of a larger, non-conservative, theory).

There are of course crudities and omissions in the account just given of how physical theories are confirmed, but none of them affect the point at issue. For instance, typically when we 'derive' a test prediction from a description of an experimental set-up, we make references to unobservables in the description of the experimental set-up, and in our derivation we implicitly use lots of theory besides the theory that we regard as under test. This undercuts the argument that the theory we regard as under test must be non-conservative; but if you take that theory and supplement it with all the bits of theory implicitly used in the derivation and all the other theory implicit in our description of the experimental set-up, the argument shows that *this* body of theory must be non-conservative for a test to be possible. And that is what we wanted: the non-conservativeness of physical theory as a whole is ultimately required for any of it to be experimentally tested.[9]

Perhaps I have unnecessarily laboured the point that science *can't* be conservative if it is to be good, while mathematics *must* be conservative to be good. But it is an important point, first because it helps undercut the influential idea of mathematics as quite continuous with science, and second because it provides a dramatic illustration of the fact that the truth of a theory (in this case, physical theory) certainly does not entail its conservativeness.

Of course the conservativeness of a theory doesn't entail its truth either. This should be clear merely from the definition of conservativeness; but in addition, it is worth noting that conflicting mathematical theories (e.g., set theory with the axiom of choice and set theory with its negation, or set theory with the continuum hypothesis and set theory with its negation) can both be conservative.[10] It seems

[8] To what extent depends, among other things, on what *other* theories are available that would lead from the given 'observational' premises to the 'observational' conclusion.

[9] For simplicity I have ignored here probabilistic scientific theories. They need not be non-conservative in quite the sense defined, but they must be non-conservative in the slightly more general sense of leading to an alteration in the probabilities of 'observational' conclusions from 'observational' premises. This alteration does not affect the basic contrast I am drawing between mathematical theories and scientific theories, because good mathematical theories are conservative in the corresponding generalized sense: addition of them to nominalistic premises never affects the probabilities of nominalistic conclusions.

[10] Cf. *Science without Numbers*, p. 15.

clear then that truth and conservativeness are two quite independent features of mathematics (if indeed truth is a feature of mathematics at all). Given any particular application of mathematics, then, it is natural to ask whether the utility of mathematics in that application is due to its conservativeness or to its truth. It is clear that an anti-realist of the sort we have been considering (one who believes that a mathematical theory that is sufficiently rich to be interesting needn't be true to be good, that it need only be conservative) is going to have to maintain that the utility of mathematics in all applications is accountable for in terms of its conservativeness alone. Let us see whether this conclusion can be maintained.

Recall that from the fact that any acceptable mathematical theory is conservative – a fact that I take it will be agreed by all once it is pointed out – it seems to follow that mathematics could be useful in only two ways. It could be useful in making it easier to infer nominalistic conclusions from nominalistic premises, and, in addition, it could be useful in being essential to the formulation of premises for some important extra-mathematical theory.

Now there are lots of examples which show that mathematics is often useful for the first reason alone. Probably the simplest examples come from the application of elementary number theory. Let N be a theory that contains the identity symbol and the usual axioms of identity, but whose other non-logical symbols are just ordinary physical object predicates and whose quantifiers range over physical objects alone. Suppose 'aardvark' is one of the predicates of N; then we can use it and the identity symbol to say in N that

1   There are exactly twenty-one aardvarks.

Similarly, with a few more physical object predicates but without mathematical objects, we can say

2   On each aardvark there are exactly three bugs.
3   Each bug is on exactly one aardvark.

and

4   There are exactly sixty-three bugs.

Suppose we want to know whether 4 follows from 1–3 in quantification theory plus identity. The answer is that it does; but an argument for 4 from 1–3 in quantification theory plus identity would be incredibly long and tedious. The introduction of number theory plus some very elementary set theory makes the argument much easier. For, *from this mathematical theory* one can prove that 1–3 hold if and only if the following hold:

1′   The cardinality of the set of aardvarks is 21.

2′   All sets in the range of the function whose domain is the set of aardvarks, and which assigns to each entity in its domain the set of bugs on that entity, have cardinality 3.

3′   The function mentioned in 2′ is 1–1 and its range forms a partition of the set of all bugs.

Similarly, the mathematical theory implies that 4 holds if and only if the following holds

4′   The cardinality of the set of all bugs is 63.

Finally, the mathematical theory enables us to prove very easily that if 1′–3′ hold then 4′ holds. Putting these together, we have a proof *using the mathematical theory* that if 1–3 holds then 4 holds too.

This is just one very simple example of how accepted mathematical theory enables us to make inferences among nominalistic claims much more easily than we could make those inferences without the mathematics. And the inferences we make in this way will be correct every time. *Prima facie* this might seem to be good evidence for the truth of accepted mathematical theory. After all, if it weren't true, invoking it in this way in arguments ought to lead sometimes from otherwise true premises to a false conclusion. But now that we've distinguished truth from conservativeness, we see that this *prima facie* plausible argument is thoroughly mistaken. The fact that in such contexts our mathematical theory never leads to error is explainable entirely by its conservativeness; its truth need not be assumed.

Several things should be noted about this account. In the first place, the anti-realist position as I construe it does not claim that 1′, 2′, etc. are equivalent in meaning to 1, 2, etc. The claim is only that the *material* equivalence of 1′ with 1, 2′ with 2, etc. *follows from standard mathematics*; and since it is part of the anti-realist position that standard mathematics is largely false, this doesn't even guarantee that 1′, 2′, etc. have the same truth value as 1, 2, etc. (In fact, they won't generally have the same truth value on the anti-realist account: e.g., 1′ will be false on most views of the semantics of non-denoting terms, however many aardvarks there are.) Any claim about the equivalence of 1 and 1′ even in truth value, much less in meaning, is gratuitous; for the utility of number theory can be explained by the less controversial assertion that the material equivalence of 1′, 2′, etc. with 1, 2, etc. follows from standard mathematics, together of course with the fact that standard mathematics is conservative.

In the second place, the above account of the application of number theory does not depend on the assumption that there are infinitely many entities. (The superficially similar logicist account, by contrast, does require this: without going into details, let me simply say that this requirement arises in the first place from the logicist's commitment to *translating* claims about numbers into claims about other entities, in an allegedly meaning-preserving way, and in the second place from a commitment to regarding the standard mathematical theorems as true.) That the above account does not depend on the assumption of infinitely many entities is fortunate for the mathematical anti-realist. For the anti-realist, there could be infinitely many entities only if there were infinitely many *physical* entities; and while it seems highly *likely* that there are infinitely many physical entities (it would seem, for instance, that there are infinitely many parts of each light wave), still it would seem quite inappropriate to rest the adequacy of number theory on any such assumptions about the physical world. It is clear that the *truth* of number theory would require infinitely many objects for its quantifiers to range over, but its *conservativeness* requires no such thing. And the point that I was making in my account of application is that conservativeness is all we need.

That is, it is all we need to account for the utility of number theory *in cases like the one described*. But surely this is an atypical application of mathematics? Consider the physicist concerned to lay down a theory of, say, electromagnetic phenomena; he or she proceeds by stating a theory involving lots of differential equations. This theory presupposes a large body of mathematics: the existence of real numbers, functions into the real numbers, differential operators and so on. Here certainly it does not seem as if we are employing mathematics to facilitate the inference of nominalistic conclusions from nominalistic premises, for where are the nominalistic premises? We would have them only if we could somehow state electromagnetic theory nominalistically, and it is *prima facie* quite difficult to see how this would be done. If this is right then mathematics is not just useful in facilitating inferences, it is *theoretically indispensable*; and it is hard to see how the theoretical indispensability of mathematics can be accounted for in terms of its conservativeness. It would appear, then, that if mathematics is theoretically indispensable, the only way to account for this is to assume that it is true.

This, I take it, is what the Quine–Putnam argument for the truth of mathematics really amounts to. Quine and Putnam tend to put their argument by saying that we need to assume the truth of mathematics to account for its utility in the non-mathematical realm, and this as we've seen is a considerable overstatement: *part* of that utility certainly can be accounted for in terms of the conservativeness of mathematics alone. I take it, however, that their main point survives this observation: if mathematics is theoretically indispensable (in physics or in other important domains) then its conservativeness is not enough to account for its utility. It may seem, then, that the emphasis in this paper on the conservativeness of mathematics doesn't ultimately take us very far in advancing the anti-realist's cause.

Actually, I think, the emphasis on the conservativeness of mathematics was not misplaced. For I believe that in the end it can be shown that mathematics is *not* theoretically indispensable; that physical theories such as electromagnetic theory can be nominalistically reformulated, and that when they are, the usual platonistic formulations of physics can then be justified using the conservativeness of mathematics. The usual platonistic formulations of physics are, on my view, rather like 1′–3′. That is, *given standard mathematics* 1′–3′ is true if 1–3 is; similarly, *given standard mathematics* the platonistic formulations of physics are true if the nominalistic ones are. If this is right, then even a nominalist is free to use the platonistic formulations of physics in drawing nominalistically statable conclusions, for by what was just said, those nominalistic conclusions would then follow from the nominalistic physics *plus standard mathematics*. That means that if standard mathematics is conservative, these conclusions would follow from the nominalistic physics alone. We see, then, that if a certain programme for how to nominalize physical theories is workable, then the role of conservativeness as a surrogate for truth is very much greater than might at first appear.

It is not my goal in the present paper to argue for the programme of showing that platonistic formulations of physical theories are simply conservative extensions of underlying nominalistic formulations; I have argued that elsewhere (at least for an important class of physical theories; and the programme of extending this to other physical theories seems to me quite a promising one).[11] Here I will consider a number of points that can be dealt with less technically.

In the first place, one might think that even if the Quine–Putnam indispensability arguments can be dealt with, there will be other considerations that favor the realist position. For instance, it is well known that Gödel advanced a number of arguments for a radically realist view of mathematics; doesn't an anti-realist have to combat these arguments too?

[11] *Science without Numbers.* The physical theories dealt with there are the classical field theories in flat space-time. In a footnote I also give a sketch of how the approach might be extended to classical field theories in curved space-time; but as I noted, this proposed extension rests on first proving a representation theorem for curved space-time, which seems possible in principle but which to my knowledge has never been done. Quantum theories of various sorts will require different methods, and as yet I have no idea how to handle them. (This is due in part, I suspect, to the fact that quite independently of nominalistic scruples, I do not really feel I understand these theories.)

In answer to this, I would like to point out that Gödel's most powerful arguments rest ultimately on the Quine–Putnam indispensability argument. Gödel (1947) argues, for instance, that even on a very narrow construal of what counts as mathematical 'data' – even, that is, if we confine the mathematical 'data' to simple number–theoretic equations – we can justify quite abstract parts of mathematics in terms of their success in explaining this data:

even disregarding the intrinsic necessity of some new axiom, and even in case it had no intrinsic necessity at all, a decision about its truth is possible also in another way, namely, inductively by studying its 'success', that is, its fruitfulness in consequences and in particular in 'verifiable' consequences, i.e., consequences demonstrable without the new axiom, whose proofs by means of the new axiom, however, are considerably simpler and easier to discover, and make it possible to condense into one proof many different proofs. The axioms for the system of real numbers, rejected by the intuitionists, have in this sense been verified to some extent owing to the fact that analytical number theory frequently allows us to prove number theoretical theorems which can subsequently be verified by elementary methods. A much higher degree of verification than that, however, is conceivable. There might exist axioms so abundant in their verifiable consequences, shedding so much light upon a whole discipline, and furnishing such powerful methods for solving given problems (and even solving them, as far as that is possible, in a constructivistic way) that quite irrespective of their intrinsic necessity they would have to be assumed at least in the same sense as any well established physical theory. (p. 521)[12]

Similarly, he remarks that even 'large cardinal axioms' in set theory have consequences that would be relevant to deciding them:

these axioms have consequences also far outside the domain of very great transfinite numbers, which are their immediate object, can be proved; each of them (as far as they are known), can under the assumption of consistency, be shown to increase the number of decidable propositions even in the field of Diophantine equations. (p. 520)[13]

Again, then, it seems like the truth of such matters can be decided (not conclusively, but with probability) by a hypothetico-deductive procedure analogous to that used in science.

Arguments of this sort have a reasonable degree of persuasiveness; that is, they make it at least fairly plausible that if we regard the so-called 'mathematical data' as true, then we ought to believe that claims about much more abstract parts of mathematics have truth value, and ought indeed to regard ourselves as having reason for believing that given such assertions have particular truth values. If this is right, then

[12] Kurt Gödel, 'What is Cantor's continuum problem?'.

[13] Ibid.

if indispensability arguments can be used to argue at least for the truth of the 'mathematical data', then indispensability arguments plus Gödelian considerations can be used to argue for quite abstract mathematical claims. In arguing for a certain large cardinal axiom, we can argue that it is plausible in light of its coherence with the indispensable part of set theory and the fact that it leads to indispensable consequences in a simpler and more unified way than the way those consequences can be derived without the axiom. But, of course, this assumes that *some* part of mathematics is indispensable. If *none* of it is indispensable, then as far as I can see, there is no reason to regard even the so-called 'mathematical data' as true, so arguments for the truth of other parts of mathematics in terms of their relationship to this data are without value.

Am I right that besides indispensability considerations, there is no reason to regard even the so-called 'mathematical data' as true? I can not go into a lengthy discussion of this here; let me simply say that every argument, not based on indispensability, that I have ever heard for regarding the 'mathematical data' as true seems to me blatantly question-begging. That is, every such argument is question-begging *against the form of anti-realism here advocated*. Certain non-indispensability arguments for regarding standard mathematics as true, such as some of the arguments considered in the opening paragraphs of this paper, have a certain plausibility to them as long as one does not have any idea what viable goal to ascribe to mathematics besides truth. But once one has an alternative account of what makes a mathematical theory good, these arguments and all other arguments besides indispensability seem totally question-begging.

Let me say a few more words about my claims regarding dispensability. In the first place, what I claim to be dispensable are mathematical entities – numbers, functions, sets and the like. I do not claim the dispensability of every kind of entity that any philosopher who calls himself a nominalist has ever objected to. In particular, I do not claim the dispensability of space-time regions in developing physical theory. Typical platonistic formulations of physics invoke space-time regions *and* numbers, functions, sets, etc.; my claim is that the numbers, functions, sets, etc. can be eliminated, but I make no such claims about the space-time regions.

The ability to maintain this position of rejecting mathematical entities while accepting space-time regions would be of only marginal interest if the grounds that have made many philosophers reluctant to postulate mathematical entities were equally good grounds for a reluctance to postulate space-time regions. But I don't think they are. Probably the main ground for suspicion about mathematical entities is the difficulty that these entities raise for the theory of knowledge and for the theory of reference or theory of belief content.[14] According to the platonist picture, the truth values of our mathematical assertions depend on facts involving platonic entities that reside in a realm outside of space-time. There are no causal connections between the entities in the platonic realm and ourselves; how then can we have any knowledge of what is going on in that realm? And perhaps more fundamentally, what could make a particular word like 'two', or a particular belief state of our brains, *stand for* or *be about* a particular one of the absolute infinity of objects in that realm? It seems as if to answer these questions one is going to have to postulate some *aphysical connection*, some *mysterious mental grasping*, between ourselves and the elements of this platonic realm.[15]

Do these considerations raise difficulties for space-time regions comparable to those that they raise for mathematical entities? No. In the first place, the difficulties just raised are not due merely to the lack of causal connections between mathematical entities and ourselves; they are due to this plus the fact that those entities exist (according to the realist) outside of space-time. Consider first the problem about knowledge. There are of course some epistemological questions that can be raised about the postulation of space-time regions, just as there are epistemological problems that can be raised about the existence of anything else; but there isn't the radical sort of epistemological problem that arises for mathematical entities, because the argument that *only a mysterious aphysical relation could account for our access to the realm of entities in question* has no plausibility here. For there are quite unproblematic physical relations, viz., spatial relations, between ourselves and space-time regions, and this gives us epistemological access to space-time regions. For instance, because of their spatial relations to us, certain space-time regions can fall within our field of vision. Of course, not *all* space-time regions are close enough to us in space and time to fall within our field of vision, but this raises no more epistemological problems for space-time regions than it raises for, say, tigers. For in addition to falling within our field of vision, space-time regions stand

[14]   There are other strong grounds for such suspicion, such as the one raised by Paul Benacerraf in 'What numbers could not be'; subsequent discussions that make clear just how difficult this problem of Benacerraf's is for the platonist are Philip Kitcher, 'The plight of the platonist' and Steve Wagner, 'Arithmetical fiction'. Another strong reason for suspicion about mathematical entities is the problem about nonstandard models, raised forcefully at the beginning of Hilary Putnam's 'Models and reality'. (Later in this article Putnam contends that the problem he raises for mathematical realism arises for realism about the physical world as well, but on this he is not very convincing.)

[15]   See Paul Benacerraf, 'Mathematical truth'.

in unproblematic relations (spatio-temporal relations) to physical objects, and this provides us with less direct observational means of knowing about them. No close analogue of this indirect observational access is available for numbers; that is why for instance we can confirm and disconfirm theories about space-time structure by observation, but can't confirm or disconfirm theories about numbers in this way. In short, what raises the really serious epistemological problems is not merely the postulation of causally inaccessible entities; rather, it is the postulation of entities that are causally inaccessible *and* can't fall within our field of vision *and* do not bear any other physical relation to us that could possibly explain how we can have reliable information about them. That is the epistemological problem that numbers and other mathematical entities raise; and space-time regions don't raise them.

Similar remarks hold for the problem of reference (by which I mean not only the problem of how words stand for things, but also the more fundamental problem of how belief states or their components stand for things). What makes the problem of reference (in this broad sense) to numbers and other mathematical entities so apparently insoluble is not merely that these entities exert no causal influence on us, but also the fact that they don't stand in other physical relations to us which could serve to explain the referential link. Space-time regions, on the other hand, do stand in unproblematic relations to us, namely spatio-temporal relations. This means, e.g., that we can point to many of them.[16] (Not to all of them, of course, but we can't point to all tigers either, e.g., we can't point to those that existed before we were born.) More importantly, it means that we can refer to certain space-time regions by means of indexicals ('here', 'now') or their mental analogues, and we can refer to many other space-time regions by means of (verbal or mental) descriptions that invoke the spatio-temporal relations that these regions bear to physical objects or to the here-and-now. It seems clear then that there is no problem of referential access to space-time regions, in the way that there is a problem about referential access to numbers.

In the preceding remarks I have not cast doubt on the assumption that space-time regions, like numbers, are causally inert; for I wanted to argue that even granting this assumption, space-time is epistemolog-

---

[16] It might be thought that there is something odd about the idea of securing reference to an object by pointing, if that object exerts no causal influence on the referrer. But I can see little motivation for this. Indeed, as Colin McGinn has pointed out ('The mechanism of reference'), reference to an ordinary physical object by means of a pointing gesture also need not depend on any causal influence of the pointed-to object on the person doing the pointing. (*Typically* there will of course be such a causal influence, but in atypical cases the causal influence may be absent and yet successful reference to a particular man by an utterance of 'that man' can occur.)

ically and referentially much less problematic than mathematical entities are. It seems to me however that the assumption that space-time is causally inert is a relic of an outdated physical theory, and that this fact makes it still more obvious that space-time regions are not epistemologically problematic in the way that mathematical entities are.

In contrast to the claim that space-time is causally inert, I claim that space-time points and regions are fully-fledged causal agents. This is a view which *can* be held on any theory that postulates space-time. For instance, instead of thinking of space-time as an inert arena in which objects like electrons reside, one can dispense with the electrons as separate entities, and regard their properties (mass-density, charge-density, etc.) as really just properties of the space-time regions that they occupy. *If we look at things in this way*, then space-time regions would appear as causal agents in the same sense that we ordinarily regard physical objects as causal agents: if their physical properties had been different, different causal consequences would have resulted.

But, it will be objected, why look at things in this way? Such a way of looking at electrons seems permissible, but not obligatory. However, when one turns one's attention from matter to fields the situation changes. For a field theory is *most naturally* construed as a theory that ascribes causal properties (electomagnetic field values, values of various gravitational tensors, etc.) to space-time points. Consequently, any theory in which talk of fields is to be taken seriously (not as mere shorthand for talk about physical objects) will be most naturally construed as a theory ascribing causal properties to space-time points; that is, as ascribing properties to space-time points in such a way that which properties are assigned causally affects the behavior of objects.

I should say incidentally that talk of 'properties' here is merely a loose manner of speaking; the nominalistic treatment of fields in my aforementioned book makes no reference to them, as well of course as making no reference to mathematical entities such as tensors. What it does do is use *causal predicates of space-time points*; and what I am saying is that this is virtually unavoidable in dealing with theories that take the notion of a field seriously.

I say 'virtually' unavoidable, for there is one way to take field theories seriously and avoid the conclusion that space-time is causally efficacious;[17] but it seems to me that this is little more than a *merely verbal* avoidance of a causally efficacious space-time, and does nothing to affect the nominalism issue that is our main concern. The way to avoid giving causal properties to space-time points is to take talk of fields in the most 'naive realist' way: to regard a field as an entity distinct from space-time, a sort of giant physical object. Where is this object?

---

[17] David Malament advocates this in his review of *Science without Numbers*.

"People slowly accustomed themselves to the idea that the physical states of space itself were the final physical reality."
PROFESSOR ALBERT EINSTEIN

Presumably it is everywhere, at all times.[18] On this view, what is causally efficacious is not space-time, but the electromagnetic field. Notice, however, that this view is little more than verbally different from the old one. For geometric relations among space-time points (betweenness, congruence, and so forth) also hold between the minimal spatio-temporal parts of the electromagnetic field; and given the omnipresence of the field, all the usual geometric assertions about space-time can now be asserted directly of the field. So what is being proposed is that there is an entity, the electromagnetic field, that causally affects physical objects by its electromagnetic properties, and that has all the geometric properties of space-time; in addition, there is another entity, space-time, that is causally inert and has the geometric properties alone. But the second of these entities is obviously gratuitous; there is no need to believe in it in addition to the other entity that has precisely the same geometric properties and also is causally efficacious. And it is a purely verbal question whether one calls the causally efficacious entity 'the electromagnetic field' or 'space-time'.[19] The important point is that in developing physical theory, the only entity that one needs besides ordinary physical objects is an entity with geometric properties and electromagnetic properties; and because these electromagnetic properties are causally efficacious,[20] the entity in question is as much a causal agent as are ordinary physical objects. One doesn't need causally inert objects like numbers in developing physical theory.

[18]   With the possible exception of the extremely small set of points at which the field in question (or each of the fields in question) takes on a value of precisely zero. Given the fact that electromagnetic etc. forces have an unlimited range, this set of points is bound to be extremely tiny. If someone were to suggest that physical fields are giant entities that occupy all of space-time except for such a tiny region, the argument of the rest of the present paragraph of the text would have to be modified slightly; I don't think however that the conclusion would ultimately be altered. (I also think that such a suggestion would have very little in the way of plausibility to recommend it.)

[19]   There is a reason for preferring the latter terminology. For consider theories involving two or more fields, say an electromagnetic field and a gravitational field. (Here I have especially in mind a gravitational field, like the scalar field of Brans and Dicke, that is not to be understood in geometric terms.) It seems more natural to suppose the existence of one object with electromagnetic and gravitational and geometric properties than to suppose the existence of two objects, one with electromagnetic and geometric properties and the other with gravitational and geometric properties. (The construal in terms of three objects, one with electromagnetic and geometric properties, one with gravitational and geometric properties and the third a causally inert object with geometric properties alone, seems less natural still.) But on the one-object view, the only natural name for the object is 'space-time': the terms 'the electromagnetic field' and 'the gravitational field' are inappropriate since the object in question has both electromagnetic and gravitational properties.

[20]   As are geometric properties, in general relativity.

The upshot, then, is

(a) that space-time regions, unlike mathematical entities, are causal agents; or if you prefer, that all talk of them can be replaced by talk of causal agents like 'the parts of the electromagnetic field'

(b) that in any case, the philosophical objections to mathematical entities do not depend merely on their acausal character, but on other features as well (such as lack of unproblematic relations to physical objects) that contribute to their 'inaccessibility'; and space-time regions do not have these features.

Either of these points alone is sufficient to undercut the objection that one cannot simultaneously dispense with mathematical objects and space-time regions, and that there isn't much value in dispensing with the former if one retains the latter.

I will now turn to one last objection to my claim that mathematical entities are dispensable. The objection is that so far in this paper I have claimed only (and in my book I have argued only) that mathematical entities are theoretically dispensable *in science*; but there is another important area of discourse that we should consider, and that is *logic*. Indeed, it might even be argued that unless we can show that mathematical entities are dispensable in logic, we haven't *really* shown that they are dispensable in science; after all, we need to use logic in developing science. And finally it might be argued that mathematical entities are clearly *not* dispensable in logic. After all, the standard semantic definition of logical consequence is in terms of *models*, which are mathematical entities; and even if one dubiously thought that it was legitimate to replace the semantic notion by a purely proof-theoretic notion, provability in a certain formal system, the fact remains that one needs abstract objects (arbitrary sequences of symbol-tokens and arbitrary sequences of such sequences) to characterize even this.

This objection raises large issues and I can not hope to adequately deal with them here; indeed, the details of my position in response to them are not entirely worked out.[21] Still, the following remarks should give the rough idea.

Let's start by considering one small part of the objection just raised, the part that says that considerations about logic might show that mathematical entities are really indispensable even to science, *despite* any nominalistic axiomatization of that science, since you need logic to draw conclusions from the nominalistic axioms. *This* part of the objection seems to me to be simply a bad argument, which turns on a failure to distinguish logic from metalogic. Logic itself is simply reasoning; the

---

[21] Note added to this volume: it is now worked out much further, in essay 3.

only entities we need to postulate here are those that occur in the premises, intermediate conclusions and final conclusions of our reasoning. To the extent that we can nominalistically axiomatize physical theories, we need no mathematical entities in our premises; since the conclusions about the world we're ultimately interested in predicting and explaining are nominalistic, we don't need mathematical entities in our final conclusions;[22] and the conservativeness of mathematics then shows that we don't need them in our intermediary conclusions. So, we don't need them *in physical theory*.

The main part of the objection remains however – we do seem to need mathematical entities in *metalogic*. To this a nominalist can respond in two kinds of ways. First, he or she could adopt a very hard-headed stand; that is, he or she could say that the claims of standard metalogic really aren't true. To give this any plausibility at all he or she would have to say about metalogic what I have said about normal mathematics – that there's nothing wrong with standard metalogic, but that the goodness of metalogic is not to be explained in terms of its truth.

Well then, how would we explain what's good about standard metalogic? A natural first answer would mimic our answer as to what's good about normal mathematics: conservativeness (in a slightly altered, but obvious, sense). That is, metalogic is good in the sense that using metalogic in the usual way enables us to simplify our arguments between non-metalogical statements, but the conclusions that we reach by these means would always be obtainable without metalogical reasoning but using only ordinary logical inference.

It is certainly true that metalogic is conservative in this sense, and that this justifies its use in simplifying arguments from non-metalogical premises to non-metalogical conclusions. Consider for instance the metalogical assertion that if a sentence B is formally derivable from (or a consequence of) sentences $A_1, \ldots, A_{n+1}$, then $A_{n+1} \supset B$ is formally derivable from (or a consequence of) sentences $A_1, \ldots, A_n$. This can be a very useful theorem to use in reasoning to a given conclusion $A_{n+1} \supset B$ from given premises $A_1, \ldots, A_n$. It is obvious that in reasoning in this way we are never led from true premises $A_1, \ldots, A_n$ to a false conclusion $A_{n+1} \supset B$, and it is obvious that the explanation of this fact requires only the assumption that metalogic is conservative and not the assumption that it is true.

The only problem is that it does not seem very likely that *all* uses of metalogic can in this way be explained in terms of conservativeness

---

[22]   Actually this second premise is not beyond challenge (as George Bealer has pointed out to me). I think that the claim can be defended against these challenges, but I will not go into these matters here.

alone. The utility of assertions about *what follows from what* can be explained in terms of conservativeness. But how about claims about *what doesn't follow from what*? Equivalently, how about claims about *consistency*? The utility of these claims doesn't seem to be to aid us in making inferences between non-metalogical assertions; rather, such claims, if we believe them, serve to *discourage us from looking* for certain sorts of arguments, and to *remove certain worries* we might have about our theories. Another sort of assertion that does the same sort of thing is an assertion of *conservativeness*; conservativeness is after all a generalized form of consistency. My assertions in this paper about the conservativeness of mathematics have certainly not had the role of enabling us to make inferences between nominalistic premises more easily. Rather, their role was to remove certain worries we might have had about our use of the systems I've claimed to be conservative.

What then of the hard-headed view about metalogic? The hard-headed view was that the claims of metalogic need not be true to be good, they need only have certain virtues that don't require truth. But what are these other virtues? Conservativeness was one of them, but it seems that to defend the hard-headed view we need to find some other virtue not requiring truth that would explain the utility of metalogical assertions of consistency, conservativeness, etc. At present at least, I am unable to see what other virtue than truth could serve this role, so at present at least I am unable to accept the hard-headed position.

As I've mentioned, however, the nominalist need not take a hard-headed stand; the alternative is to adopt an attitude of belief toward our normal metalogical claims, but to try to account for these metalogical claims without introducing mathematical entities. I think that this or something very close to this can be done, but it requires something that is not required in nominalizing physical theories: it requires an expansion of our logic to include the modal operator 'it is logically possible that'. Of course, a nominalist who uses such a logical operator should not try to explain it in terms of 'possible worlds'; for 'possible worlds' are abstract entities that are no less problematic as far as epistemology and reference are concerned than are mathematical entities.[23] What the

---

[23] In calling 'possible worlds' abstract I ignore David Lewis' view (*Counterfactuals*, section 4.1) that the actual universe is just a 4-dimensional slice of a broader reality, and that the other 'possible worlds' are just other 4-dimensional slices of this same broader reality and hence are no more abstract than the actual universe. This view seems to me exceptionally implausible; in addition, it can have little appeal for the nominalist, for even if these other possible worlds are not abstract entities, the problems posed by knowledge of them and reference to them seem just as intractable as the analogous problems for numbers.

nominalist who wants to use such a modal operator should say is that there is no need to explain this operator in other terms, any more than there is a need to explain the negation operator or the existential quantifier in other terms; instead, we should regard the modal operator as simply a logical primitive, one that we come to understand not by defining it but in whatever way we come to understand the other logical primitives.[24]

Is this a legitimate attitude? It seems to me that it would be highly improper to use a logical possibility operator (or any other sort of possibility operator) in nominalizing physics; physics, as I see it, has as its only concern the description of the actual features of the actual world, and I don't think it appropriate to invoke facts about what is possible in doing this. (This is not an argument but merely an expression of taste.)[25] It seems to me, however, that in the realm of metalogic the situation is very different. Possibility seems intimately connected with logic in a way that it is not intimately connected with physics: logic is the science of the possible. Indeed, the usual platonistic treatments of metalogical notions seem artificial precisely because they avoid bringing in possibility. Isn't it more natural to characterize the *provability* of a sentence in terms of the *possible* existence of a string of symbols that meets the conditions of being a proof of it, than in terms of the *actual* existence of an abstract sequence of abstract analogues of those symbols? And isn't the semantic consistency of the theory of discrete linear orderings with no last element more naturally thought of in terms of the *possible* existence of entities that are ordered in the way this theory says, rather than in terms of the *actual* existence of an ordered pair whose first member is an infinite set and whose second member is a subset of the Cartesian product of that set with itself? I would have thought so.[26]

It seems to me that the widespread acceptance by philosophers of the usual set-theoretic handling of metalogical notions is due in part to the idea that the *prima facie* more natural explanation in terms of possibility would in the end turn out at least as bizarre as the set-theoretic explanation, once possibility is explained in terms of possible worlds.

---

[24] For a defence of this view, see Colin McGinn, 'Modal reality'. However, unlike McGinn, I would confine my acceptance of a primitive possibility operator to an operator of *logical* possibility.

[25] Note added to this volume: for an argument, see essay 6, sections 6 and 7.

[26] It might indeed be argued (after the fashion of David Lewis, op. cit.) that set-theoretical explications of logical consequence etc. are incorrect, on the grounds that they allow as consistent various things that aren't, e.g., 'There are married bachelors'. Such an argument opens several rather large cans of worms and I am not sure whether or not it is ultimately correct.

But if we refrain from explaining possibility in terms of possible worlds, if we regard the modal operator as simply a primitive part of logic which no more requires explanation in terms of possible worlds than the existential quantifier requires explanation in terms of functions from sets into truth values, then this argument fails and the treatment of metalogical notions in terms of possibility does indeed remain the more intuitively correct.

I must admit that I do not entirely welcome the introduction of the notion of possibility even into metalogic; it seems much less bad than either the introduction of this notion into physics or the introduction of pieces of platonic protoplasm with no causal connection to us or to anything we observe and existing outside of space-time, but still it is something I would prefer to avoid if possible. And I am not ultimately sure that it isn't possible; the hard-headed position that I described a few pages back goes an important part of the way toward working, and it would be pleasant to think that by coming up with another virtue distinct from both conservativeness and truth we could explain what's so good about standard metalogic without assuming it true. But until we find a way to do this, it seems to me that acceptance of a primitive logical possibility operator in dealing with metalogic is the best position available.

Postscript

1   I somewhat regret the discussion of truth at the start of this essay. My main point was simply that acceptance of a standard conception of truth for mathematics doesn't itself require the existence of mathematical entities: it requires this only given the additional assumption that something like the usual mathematics is true (an assumption I deny). But the way that I put things has suggested to some that it is only if we accept a Tarski-like theory of truth that mathematics requires the existence of mathematical entities; and this has led to the misconception that without a Tarski-like account of truth, mathematics would raise no epistemological problems. (See Tait 1986. In conversations I have even heard it claimed that epistemological problems arise only if we accept a modified Tarskian theory of truth of the sort discussed in Field 1972, together with the non-trivial theories of reference for the primitives of the language that I advocated in that paper.) In fact, of course, mathematics itself entails the existence of mathematical entities, without any help from a notion of truth: mathematics itself entails, for instance, that there are prime numbers greater than 1,000. It is the fact that mathematics implies the existence of mathematical entities that creates the most fundamental epistemological problems for believing mathematics. Talk of truth is best left out of the discussion.

2   The discussion of the aardvarks example is over-simplified, though the over-simplification does not affect any of the points I was trying to make with it. The points I was trying to make were (a) that we use the theory of natural numbers in deciding which inferences in first order logic are legitimate, and that this use of number theory is important in that without it a decision about the legitimacy of many such inferences would be unmanageable; but (b) that it is *legitimate* to use number theory in this way even if we don't believe that there are natural numbers, because of the fact that number theory is conservative. These points are correct. But in my discussion of (a), I over-simplified: the account I presented of how to deduce 4 from 1–3 by way of number theory is really not much more manageable on its own than is the direct deduction of 4 from 1–3 in logic. (The problem is in the proof of the equivalence of 1 with 1′, etc.) To get an accurate account of how we deduce 4 from 1–3 in practice, one would need not only number theory but also metalogic (model theory or proof theory), and metalogic like number theory postulates mathematical entities. But this fact does not affect the legitimacy of making the deduction without literally believing in mathematical entities: for as I say later in the paper (and elaborate in essay 3), metalogical theories are like number theory in being conservative.

3   I remark that just as normal mathematics is conservative, so too metalogic is conservative 'in a slightly altered, but obvious, sense'. In the case of proof theory, the sense is this: whenever 'There is a derivation of B from Γ' follows from proof theory, B follows from Γ. See the discussion of modal soundness principles in essay 3.

4   The 'soft' stand on metalogic now seems to me clearly right on the crucial point: we do need to utilize a notion of logical possibility in dealing with metalogical notions. (This is discussed in some detail in essay 3.) Quite a few of the objections that have been raised against *Science without Numbers* no longer arise once metalogical notions are construed modally and recognized as perfectly legitimate. For instance, the unfavourable comparison of my programme with Hilbert's in Detlefsen (1986) would not have been possible had I given the modal construal of metalogic in my book. Nor would Malament's complaint (1982) that my nominalism is too austere to say that there are models of the Klein–Gordon equation in which the field is non-constant, or that the theory of this field is deterministic: the effect of these can be got by modal assertions (involving a purely logical modality) in which the nominalized Klein–Gordon theory that Malament alludes to occurs as a component.

3

Is mathematical knowledge just logical knowledge?*

Logicism is the claim that mathematics is part of logic. This claim flies in the face of Kant's denial that mathematics is analytic, that is, true by logic and definitions alone; and it seems to me that if mathematics is taken at face value, Kant is surely right.

The reason for this assessment is that mathematics, if taken at face value, makes existential assertions: it asserts for instance that there exist prime numbers greater than a million, and therefore that there exist numbers. Indeed, even Kant's example '5 + 7 = 12' makes an existential assertion, if understood in the usual way: it asserts not only that *if* there are x, y and z such that $x = 5$ and $y = 7$ and $z = 12$ *then* $x + y = z$, but the further existential claim that there *are* such x, y and z. So to argue against the idea that mathematics, if taken at face value, is true by logic and definitions alone, we only need argue that you can't get existential assertions out of logic and definitions alone.

And Kant did provide such an argument (though not in his discussion of mathematics). Anselm, Descartes and others had argued that the

* I would like to thank Tony Anderson, Paul Benacerraf, Charles Chihara, Geoff Joseph, Janet Levin, David Lewis, Penelope Maddy, Colin McGinn, Bill Tait, Richard Warner and an anonymous referee, for helpful comments that have affected the final version. Much of the research was carried out under a summer research grant from the National Science Foundation (SES-8205264).

Note to this volume: I have done some rewriting of this paper. Apart from section 4, the only substantive revision is in the last paragraph of fn. 15. Section 4 has been considerably rewritten, in an effort to clarify my position and remove some minor errors. The main change is that in the original version of the paper I introduced a 'broad conception of standard mathematics', replete with substitutional quantifiers and an ω-rule for them; but I have come to see that my point could be made more clearly without this. There are also changes in my remarks on nominalistic proof theory. Not everything that I would change if I were writing the paper today has been changed: see the postscript for additional remarks.

existence of God is a matter of logic, of conceptual necessity: that it follows from the very concept of God that God exists. Kant argued that this can't possibly be correct, for logic (and logic together with definitions) can never categorically assert the existence of anything. Kant's argument for this principle is that contradictions usually stem from postulating one or more objects and making various assumptions about the postulated object or objects that are mutually inconsistent: for example, postulating a triangle, and then saying something else that implies that it has more than three sides. But there is never a contradiciton if we reject the triangle – there is nothing there about which we have made contradictory assumptions. And, to quote Kant, 'the same holds true of the concept of an absolutely necessary being. If its existence is rejected, we reject the thing itself with all its predicates; and no question of a contradiction can then arise.'[1] He sums it up by saying 'I can not form the least concept of a thing which, should it be rejected with all its predicates, leaves behind a contradiction.'[2] I think that this argument is rather persuasive. If it is correct, it cannot be contradictory to deny the existence of God; and it cannot be contradictory to deny the existence of numbers either, for they don't have the mysterious power of leaving behind a contradiction when their existence is rejected any more than God does.

One can quibble with this argument of Kant's for the principle that logic and definitions alone imply no existence assertions; nevertheless, the principle itself is a very compelling one. Perhaps when a person denies the existence of God or of numbers, what the person is saying is false or even 'metaphysically impossible' (whatever that means); but it is not itself a logical contradiction in any normal sense of 'logical contradiction'. Moreover, there is good reason not to depart from the normal sense of 'logic' by counting existence assertions as part of logic: doing so would tend to mask the fact that there is a substantive epistemological question as to how it is possible to have knowledge of the entities in question (God, numbers, etc.).[3]

---

[1] *Critique of Pure Reason*, B622–3.

[2] *Ibid.*, B623–4.

[3] Admittedly, there are also questions about how it is possible to know logical truths like 'If snow exists then snow is white or snow is not white'; but such questions seem much less gripping than questions about how we can know the existence of specific kinds of entities, and it seems very unlikely that any reasonable answer to the question of how logic is known would bring with it an answer to the question of how the existence of God or of numbers or of any other specific sorts of entities is known. See for instance Paul Benacerraf, 'Mathematical truth' for a discussion of epistemological difficulties that our apparent knowledge of the existence of mathematical entities raises and which don't seem to be raised by straightforwardly logical knowledge. (In essay 2 I discuss this further and argue that contrary to what is sometimes claimed, the epistemological difficulties that

So mathematics, taken at face value, can not be reduced to anything reasonably called logic. In a sense, of course, the classical logicists did take mathematics at other than face value: they held that though it at first blush appears to concern specifically mathematical entities like natural numbers, real numbers and tensors, it can really all be shown to be part of the theory of properties (or the theory of propositional functions, or the theory of extensions of concepts, or whatever). But the theory of properties, or propositional functions, or whatever to which the classical logicists hoped to reduce mathematics asserted the existence of a vast array of properties (or propositional functions), so the problem of how this can reasonably be regarded as logic recurs. (It is, indeed, a problem which eventually led one famous logicist, Frege, to abandon his logicism: 'it seems that [logic] alone cannot yield us any objects . . . [So] probably on its own the logical source of knowledge cannot yield numbers'.[4] If one is going to retain logicism (in conjunction with the Kantian requirement on logic that I have advocated), one is going to have to provide an interpretation of mathematics according to which mathematics does not really make existential assertions despite all appearances to the contrary. I do not regard the prospects of doing this in a plausible way as at all promising, so I will not be defending logicism.

Still, I think that the idea that mathematical knowledge is just logical knowledge is largely correct, for I want to maintain what might be called a *deflationist* position about mathematical knowledge. That is, I want to say that what separates a person who knows a lot of mathematics from a person who knows only a little mathematics is *not* that the former knows many and the latter knows few of such claims as those that mathematicians commonly provide proofs of (i.e., of those claims, such as that there are prime numbers greater than a million, which I have claimed to be non-logical). Rather, insofar as what separates them is knowledge at all,[5] it is knowledge of various different sorts. Some of the knowledge that separates them is empirical knowledge (e.g., about

---

Benacerraf discusses do not depend on assuming a causal theory of knowledge.) It would hardly be a solution to the problem that Benacerraf raises to say that we know that numbers exist because *logic guarantees* that they exist.

[4] 'A new attempt at a foundation of arithmetic', pp. 278–9.

[5] Another thing that separates those who know lots of mathematics from those who know only a little isn't knowledge at all strictly speaking, it is *ability* ('knowledge-how' as opposed to 'knowledge-that'): the ability to prove mathematical theorems, the ability to see the relevance of mathematical theorems to practical matters, and so forth. But only to the extent that the possession of such abilities depends on the possession of knowledge-that does the possession of such abilities raise epistemological problems; that is why in the text I have focused only on the *knowledge* (knowledge-that) which separates those who know lots of mathematics from those who know little.

what other mathematicians accept and what they use as axioms). Putting empirical knowledge aside, my claim is that the rest of the knowledge that separates those who know lots of mathematics from those who know only a little is knowledge of a purely logical sort – even on the Kantian criterion of logic according to which logic can make no existential commitments.

The epistemological advantages of such a view are obvious: it obviates the need of postulating mathematical knowledge that is not logical and hence that is presumably synthetic *a priori*; and (putting questions of *a prioricity* or *a posterioricity* aside) it obviates the need for postulating epistemological access to a special realm of mathematical entities. Nonetheless, it is not at all obvious how any such deflationist view is to be worked out in detail, or how plausible it can be made. This paper is an attempt to survey some of the main problems that must be overcome in defending a deflationist view and to suggest ways of dealing with them.

1

The crudest attempt to state a deflationist position would be to say that all mathematical knowledge is really just knowledge that certain mathematical claims follow from certain other mathematical claims and bodies of claims: we can know, for instance, that the claim that there are primes greater than a million follows from the usual axioms of number theory. (This form of deflationism is reminiscent of, but importantly different from, what is sometimes called 'deductivism' or 'if-thenism'. Some of the differences will be discussed near the end of the essay.) This crude form of deflationism is difficult to believe: for in addition to knowing that certain claims follow from certain bodies of other mathematical claims, don't we also know the consistency of some of those bodies of mathematical claims? For instance, don't we know the consistency of various mathematical axiom systems? If one were to take the crude form of deflationism seriously, one would have to say that we can't really know that an axiom system is consistent: we can know only that the consistency of one axiom system follows from the axioms of another system which itself can't be known to be consistent (though of course its consistency can be known to follow from still a third axiom system). This strikes me as extremely implausible.

Fortunately, there is no need for a deflationist to be a crude deflationist, and so no need for him or her to take this line on consistency. For it would seem that knowledge that a certain axiom system is consistent (i.e., knowledge that some claim of the form 'p & -p' *doesn't follow* from the system) is every bit as much *logical* knowledge as is knowledge that a certain claim *does follow* from the system. The implausibility of the crude form of deflationism lies in its being forced to try to explain apparent knowledge of *what doesn't follow* in terms of knowledge of *what does follow*. But there is no point in trying to do this if both have equal claims to count as logical knowledge.[6]

A less crude form of deflationism, then, is the view that the only knowledge that differentiates a person who knows lots of mathematics from a person who knows only a little (aside from empirical knowledge of various sorts, such as the sort mentioned earlier) is

(i) knowledge that certain mathematical claims follow from certain other mathematical claims or bodies of claims,

(ii) knowledge of the consistency of certain mathematical claims or bodies of claims

and other knowledge of a basically similar sort; and that all this knowledge is logical.

Unfortunately, however, there is a powerful objection to this less crude form of deflationism (and to the cruder form as well); and that is that the knowledge cited in (i) and (ii) of the previous paragraph is *not* logical knowledge. For instance, the knowledge in (i) – that certain mathematical claims follow from certain others – isn't *logical* knowledge (i.e., knowledge of logical truths); it is *metalogical* knowledge, for it is knowledge *about the relation of logical consequence*. Now, the relation of logical consequence can be understood in either of two ways. It can be understood semantically; in that case, to say that A is a logical consequence of T is to say that in all models in which T is true, A is true as well. But so understood, the knowledge that A is a consequence of T is knowledge about *all models*. Models are mathematical entities, so knowledge about logical consequence understood semantically is a special sort of mathematical knowledge. It is not knowledge of a logical truth, such as 'If snow exists, then snow is white or snow is not white.' So much for the semantic construal of logical consequence; but how about the syntactic construal, on which to say that A is a logical consequence of T is to say that there is a formal derivation of A from T (according to the rules of some specific formal system)? Clearly this is no better, for then knowledge that A is a consequence of T in the syntactic sense is knowledge *of the existence of formal derivations*. This cannot be logical knowledge: logic can't assert the existence of formal

---

[6] As we will see in the next section, there are grounds which could lead some people (though they won't lead me) to deny that both have equal claims to count as logical knowledge.

derivations any more than it can assert the existence of God. Indeed, it is mathematical knowledge, for formal derivations in the intended sense are the abstract objects dealt with in the mathematical theory of proof. (They aren't simply strings of symbols on paper, for A can be a consequence of T without there being a piece of paper on which someone has written a formal derivation of A from T.)

I have stated the objection as an objection to the claim that the kind of knowledge mentioned in (i) is logical knowledge; but clearly the objection holds equally well for the kind of knowledge mentioned in (ii). One might be tempted to conclude that the idea that there is no mathematical knowledge over and above logical knowledge is simply a mistake.

I want to resist this conclusion, and to see how to do this, it is worth noting that even independently of the fact that you seem to need mathematical entities to define logical consequence and consistency, there is something else unintuitive about the idea that the only mathematical knowledge there is is strictly speaking of form (i) or (ii) or something similar. For the knowledge cited in (i) and (ii) is metalinguistic; are we really to hold that all mathematical knowledge is metalinguistic? Indeed, doesn't the metalinguistic fact that one sentence follows from another depend on the fact that certain words appearing in these sentences are used as logical words by speakers of the language in question? If so, knowledge of what follows from what has a contingent element that the mathematical knowledge we were trying to convey presumably lacks. This then is another (admittedly less compelling) objection to the version of deflationism under discussion. And it seems clear that we can get around both objections simultaneously if we can find a way to 'semantically descend', that is, to state the sort of mathematical knowledge that the deflationist wants to focus on without going metalinguistic.

How this is to be done is clearest in the case of finite bodies of mathematical claims, and for the moment I will confine my attention to them. If A is the conjunction of all members of a body T of mathematical claims (e.g., the conjunction of all the axioms of a finitely axiomatized theory), then instead of saying that we have mathematical knowledge that this theory is consistent, why not simply say

(ii′) we know that $\Diamond$A

(where the modal operator '$\Diamond$' is to be read 'it is logically possible that' or 'it is logically consistent that'). Here the claims in T are used, not mentioned, so the contingency objection doesn't apply. And because they are used, not mentioned, the symbol '$\Diamond$' cannot be understood as a predicate that needs explanation in set-theoretic or proof-theoretic terms; it must be understood as an operator, and indeed an operator that is widely regarded as an operator of logic; consequently the earlier objection doesn't apply either.

The point I have made for (ii) applies of course to (i) as well. That is, instead of saying that the claim B follows from the body of claims whose conjunction is A, why not say

(i′) we know that $\Box(A \supset B)$,

where '$\Box$' ('it is logically true that') is of course defined as '$-\Diamond-$'.

So our third version of deflationism (the final version, apart from a slight alteration designed to handle theories that are not finitely axiomatized) is that what differentiates a person with lots of mathematical knowledge from a person with only a little (apart from differences in abilities (cf. footnote 5) and in empirical knowledge) is that the former but not the latter has lots of knowledge of the type (i′) and (ii′) and other knowledge of a basically similar sort. (One of the things that this last 'hedge' clause covers is modal knowledge not either of the form '$\Diamond A$' or '$\Box A$'; for instance, the conditional knowledge that *if* it is consistent that A *then* it is consistent that B seems like knowledge that a deflationist could perfectly well allow.)[7]

In moving from (i) and (ii) to (i′) and (ii′), I of course have to accept the idea that the notion of logical possibility is an acceptable notion,

---

[7] It should be noted that the modal knowledge which deflationism allows is knowledge of purely logical possibility – deflationism does not allow knowledge of mathematical possibility in any interesting sense. This makes deflationism very different from the viewpoint that Hilary Putnam calls 'mathematics as modal logic' in 'What is mathematical truth?'. According to 'mathematics as modal logic',

> the mathematician . . . makes no existential assertions at all. What he asserts is that certain things are possible and certain things are impossible – *in a strong and uniquely mathematical sense of 'possible' and 'impossible'*. (p. 70, my italics.)

Putnam claims that despite its making no existential assertions, mathematics as modal logic really states the same facts as does mathematics on its platonistic interpretation. I think that there is considerable plausibility in Putnam's claim. The reason is that in the 'strong and uniquely mathematical sense of "possible"', 'it is possible that A' is an object-level analogue of 'A is consistent with any true mathematical theory'; something very akin to mathematical truth (and therefore to mathematical existence) is being sneaked into Putnam's possibility operator. Putnam's position apparently is that if you take ordinary set theory S and some nonstandard alternative S′ that is inconsistent with S (but internally consistent), then S is mathematically possible but S′ isn't. The deflationist viewpoint is different: S and S′ are on par in that $\Diamond Ax_S$ and $\Diamond Ax_{S'}$. Of course, S may be more useful than S′ for various purposes, but if so, this requires an explanation. (See sections 3 and 4 of this essay.) A deflationist cannot regard it as an acceptable explanation to say that S describes a mathematical possibility and S′ does not. I believe that Putnam's view is that such an explanation *would* be acceptable and, indeed, would be the only explanation possible.

and also I must accept that it really is a part of logic and not something that must be explained in terms of *entities* (e.g., models, formal derivations or possible worlds). In addition, in accepting that (ii′) counts as logical knowledge (for suitable assertions A), I have to interpret the idea of logical knowledge (and logical truth) in a way broader than current orthodoxy permits. This I will now explain.

2

Consider the claim

$$(1) \quad \Diamond \exists x \exists y \ (x \neq y),$$

which says that it is logically possible that there be at least two objects in the universe. Is this a truth of logic? I think the natural answer is 'yes', and most of the people I have asked agree. Nonetheless, on the usual approach to defining logical truth for modal logic (Kripke's approach),[8] (1) does not come out logically true. Indeed, it is a curious feature of Kripke's approach to defining logical truth for modal logic that *no* sentence of the form '$\Diamond$A' is logically true, except in the trivial case where A itself (and hence $\Box$A) is logically true. To me, this seems quite unmotivated. It may be countered that while it is perhaps initially natural to regard (1) as a logical truth, Kripke's model-theoretic definition of logical truth for modal logic is also quite intuitive, and it is a consequence of this model theory that (1) not be logically true. My reply is that there is an alternative model-theoretic definition of logical truth for modal logic that is simpler and I think more natural than Kripke's, and which is much closer than Kripke's to the model-theoretic definition of logical truth for first order logic (for instance, in involving no reference to 'possible worlds' or to any other entities not used in the model theory for first order logic); and according to this alternative model-theoretic definition of logical truth, (1) comes out logically true. (The basic idea of this alternative method of defining logical truth has occurred to quite a few people, starting with Carnap. I describe the version of it that I favour in an appendix).

Indeed, it is a consequence of the non-Kripkean approach to defining logical truth for modal logic that any assertion of the form '$\Diamond$A' that is true is logically true, and that any assertion of the form '$\Diamond$A' that is false is logically false. It is essential to the plausibility of this that '$\Diamond$' be read 'it is *logically* possible that' – it is *logical* possibility (not mathematical possibility or 'metaphysical possibility') that we are concerned to give the logic of. (See footnote 7.) Exactly what the force

---

[8]   'Semantical considerations on modal logic'.

of 'logically possible' is depends on further stipulation: it depends on what one takes non-modal logic to include. Some philosophers (e.g., Carnap) regard the non-modal logic to which we are adding '$\Diamond$' as including not only first order logic properly so called, but also 'meaning postulates' specifying 'logical' relations among predicates. Consequently, a non-modal sentence such as

(2) $\exists x$ (x is a bachelor & x is married)

would count as *logically* false for Carnap, and as a result,

(3) $\Diamond \exists x$ (x is a bachelor & x is married)

also comes out logically false. But I prefer not to follow Carnap in taking meaning relations among predicates to be part of logic. My preference is not based on any firm doctrines about analyticity; indeed, my preference is partly based on a desire to remain neutral about such issues. Mostly, however, my preference is based on simplicity: it is simpler to develop a basic modal logic that takes no account of meaning relations among predicates; once one has such a logic, it is easy to obtain from it a derivative logic that takes into account any relations among predicates that one cares to regard as meaning-postulates, if one so desires. If one adopts this strategy – and I shall – then (2) is not *logically* false; it is *logically* consistent that there be married bachelors (even though it may not be consistent with meaning-postulates that there be married bachelors). Now, if it is *logically* consistent that there be married bachelors, and '$\Diamond$' is read as an operator meaning 'it is *logically* consistent that', then (3) comes out true. Indeed, any sentence of the form '$\Diamond \exists x$ (Px & Qx)', where P and Q are atomic, comes out true; what else could you expect if meaning relations among predicates are not taken into account? Moreover, it seems as if it ought to be part of the *logic* of the logical consistency operator '$\Diamond$' that sentences of the form '$\Diamond \exists x$ (Px & Qx)' are true. That is,

(4) $\Diamond \exists x$ (x is red & x is round)

should be not only true but logically true; and similarly for (3). The view that (3) shouldn't be logically true, indeed shouldn't even be true, results from giving to '$\Diamond$' a sense not intended.[9]

---

[9] As remarked above, there need be no doctrinal difference between the view I have advanced, according to which (3) counts as logically true, and Carnap's view, according to which a sentence typographically like (3) counts as logically false. For we can represent the Carnapian view within the view I have advanced by introducing Carnapian notions by definition into both the object language and the metalanguage. Thus 'is C-logically true' and 'is C-logically false' are to mean 'follows from . . .' and 'is inconsistent with . . .', where the blanks are to be filled by the 'meaning postulates' for English; and

cont. over page

I hope this gives some idea of (and some motivation for) the view of modality and of the logic of modality that I will be presupposing. For a bit more detail, see the appendix.

Let us return to the issue of mathematical knowledge, and in particular to the version of deflationism arrived at in section 1. Part of the position arrived at there was (a) that mathematicians sometimes know things of form $\Diamond AX$, where AX is a conjunction of axioms of a theory; and (b) that this knowledge is logical knowledge. Now a minimum condition for (b) to hold is that $\Diamond AX$ must count as a logical truth. AX itself won't be logically true, if it is a conjunction of axioms of a typical mathematical theory; so in order to adhere to the version of deflationism put forth in section 1 we clearly have to adhere to a non-Kripkean conception of logical truth according to which some non-trivial assertions of possibility are part of logic. The non-Kripkean conception of logical truth sketched above (or if you prefer, the more fully Carnapian variant sketched in footnote 9) will do. Indeed, they have the feature that the modal assertion $\Diamond AX$ will be logically true if it is true at all. So there is no danger on these conceptions that we might know that $\Diamond AX$ and what we know not be logically true: if we know it, it's true, and so it's logically true.

Does this show that our knowledge that $\Diamond AX$ is logical knowledge? Not by itself – for it might be claimed that though $\Diamond AX$ is logically true, it is known by non-logical means.

There is a more interesting and a less interesting version of this claim. The less interesting version points out that much of our knowledge of possibility is to some extent inductive. For instance, our knowledge that $\Diamond AX_{NBG}$ (where NBG is von Neumann-Bernays-Gödel set theory and AX_{NBG} is the conjunction of all its axioms) seems to be based in part on the fact that we have been unable to find any inconsistency in NBG. And, it can be claimed, this inductive element in our knowledge precludes that knowledge from being logical. Now, even this less interesting version of the claim that our knowledge of the form $\Diamond AX$ is non-logical raises some interesting issues about the nature of logic,

---

'$\Diamond_C A$' is to mean '$\Diamond(A \& \ldots)$.' Then though (3) is still logically true (and hence C-logically true), still

(3_C)　$\Diamond_C \exists x$ (x is a bachelor & x is married)

is C-logically false (and, indeed, logically false). Moreover, we may if we like agree with Carnap that it is C-logical truth and '$\Diamond_C$' that are the philosophically more important notions. Whether or not one agrees with that philosophical claim, the procedure of focusing first on logical truth and on '$\Diamond$' is of quite considerable technical convenience. This will be evident, for instance, when we come to formulate the Conditional Possibility Principle later in this section.

issues that may be relevant to the precise wording of the deflationist's claim. But there is no need to go into such matters, for it seems quite clear that the basic idea of deflationism cannot be undercut by pointing out that much of our knowledge of possibility has a partly inductive character. The basic idea of deflationism is that one is to avoid postulating knowledge of a realm of mathematical entities, and that one is to do this by saying that ordinary mathematical claims are not known to be true. The deflationist holds that what separates those who know a lot of mathematics from those who know only a little is various sorts of knowledge and abilities, none of which give rise to the philosophical problems that knowledge of a realm of mathematical entities gives rise to (or is commonly thought to give rise to). A good deal of the knowledge that separates those with lots of mathematical knowledge from those with only a little is empirical. (I mentioned this earlier, and will discuss it more fully near the end of the essay.) Other of this knowledge is, let us suppose, *straightforwardly* logical in that it involves no inductive elements. And other of this knowledge is knowledge of logical truths by partly inductive means. Perhaps the fact that this latter knowledge is partly inductive keeps it from being logical, and perhaps not. Perhaps it makes the knowledge empirical, perhaps not. I would incline toward answering both of these questions in the negative; but however they are answered, the fact that some of our knowledge of logical truths is partly inductive does not in any way support the claim that it is based on knowledge of a special realm of abstract entities or on knowledge of the truth of ordinary mathematics. Because of this, the fact that some of our knowledge of logical truths is in part inductive can't be used to argue against the essentials of the deflationist position.

As I've remarked, there is also a more interesting version of the claim that though $\Diamond$AX is logically true, it is known by non-logical means – a version which, if true, would genuinely count against deflationism. Consider what Frege said about knowledge of the consistency of mathematical theories:

Strictly, of course, we can only establish that a concept is free from contradiction by first producing something that falls under it. (p. 106e)[10]

Obviously this is not literally correct – we can establish that the concept 'horse with wings' is free from contradiction without producing a horse with wings – but the position can be weakened without totally altering its spirit. The weakened version of Frege's claim grants that there is knowledge of possibility that does not arise from knowledge of actuality, but which arises instead from reflection on the logical form of concepts.

---

[10] §95 of *The Foundations of Arithmetic*.

But, it maintains, all such knowledge of possibility is conditional: one cannot attain categorical knowledge of possibility by this means alone. Rather, categorical knowledge of possibility can only be obtained either directly from knowledge of actuality, or indirectly, that is, from direct categorical knowledge of possibility in conjunction with conditional knowledge of possibility and other logical knowledge. So, for instance, reflection on the common logical form of 'horses with wings' and 'animals with tails' yields the conditional knowledge that *if* it is logically possible that there are animals with tails, *then* it is logically possible that there are horses with wings.[11] This knowledge together with the knowledge that there actually are animals with tails then yields knowledge that it is logically possible that there be horses with wings. The Fregean position is that all knowledge of possibility arises by some such means. (Of course, the knowledge of actuality on which knowledge of possibility is ultimately based may, on Frege's view, be *a priori*.)

If this Fregean position about knowledge of possibility were correct, then deflationism would be in deep trouble. For presumably we know (or at least have good reason to believe) the claim $\Diamond AX_{NBG}$ and the claim $\Diamond AX_R$ where R is the theory of real numbers; but according to deflationism, we do not know (or have good reason to believe) the claims AX_{NBG} or AX_R themselves since they assert the existence of mathematical entities. If Frege were right, then our knowledge that $\Diamond AX_{NBG}$ and $\Diamond AX_R$ would have to be based on conditional knowledge of possibility that arises by reflection on logical form, together with other logical knowledge plus knowledge of actuality. Now, one principle that I think even a Fregean would grant is that if ϕ is non-modal and ψ is a generalized substitution instance of it (i.e., is obtained from ϕ either by substituting formulas for non-logical predicates or by uniformly restricting the ranges of all quantifiers and free variables or both,[12] then we can know that *if* $\Diamond\psi$ *then* $\Diamond\phi$ by reflection on logical form alone.[13]

---

[11]　Here and throughout the rest of this section, the fact that we have not included meaning relations among predicates as part of logic pays off.

[12]　The restriction of quantifiers and free variables is to be by a formula $D(x_1)$ which may contain other free variables besides x_1; the formula $A_p(x_1,\ldots x_k)$ to be substituted for the k-place predicate p may likewise contain other free variables. The restriction on quantifiers and free variables is to be made only on the quantifiers and free variables of the original formula, not on any new ones introduced in an A_p or in D. (To be more formal: before performing the general substitution in a formula B, replace all bound variables in B or in D or in an A_p that occur (free or bound) in any other of the formulas by new variables. Then for any sub-formula X of B, associate an X^* as follows: if X is $p(v_1,\ldots,v_k)$, let X^* be $A_p(v_1,\ldots,v_k)$; if X is -Y, or $Y \supset Z$, let X^* be $-Y^*$, or $Y^* \supset Z^*$; if X is $\forall v Y(v)$, let X^* be $\forall v(D(v) \supset Y^*(v))$. Finally, the generalized substitution instance $\hat{B}$ of B is $D(y_1)\ \&\ \ldots\ \&\ D(y_n)\ \&\ B^*$, where $y_1,\ldots y_n$ are the variables free *in B*.) The possibility of $\hat{B}$ is in effect the possibility of its existential quantification, both with respect to the variables free in B (now restricted by D) and with respect to any other

If this Conditional Possibility Principle is granted to the Fregean, then by embedding real number theory R in set theory NBG, the Fregean can admit that we can know that

(i) If $\Diamond AX_{NBG}$ then $\Diamond AX_R$.

(For if R' is the set-theoretic assertion to which the conjunction of the axioms of real numbers 'reduce', then the knowledge in (i) arises from the knowledge that

(ii) $\Box(AX_{NBG} \supset R')$

together with the knowledge that

(iii) if $\Diamond R'$ then $\Diamond AX_R$;

(ii) is knowledge of necessity rather than of possibility,[14] and on the Fregean view, this is unproblematic; and (iii) is knowledge that results by the Conditional Possibility Principle just given.) But it is essential to this example that NBG be at least as rich as R. From the Fregean standpoint, any reason to believe that $\Diamond AX_{NBG}$ has to rest either on a reason to believe $\Diamond T$ for some *richer* theory T or else on a reason to believe AX_{NBG}. The deflationist cannot allow that there is any reason to believe either AX_{NBG} or any other mathematical theory. It is *compatible* with deflationism that there is an empirical theory T richer than NBG which can reasonably be believed; but (a) it is hard to believe that there is any plausible empirical theory in which NBG can be embedded, and (b) it is totally implausible that our reasons for believing that $\Diamond AX_{NBG}$ should rest entirely on reasons to believe any specific empirical theory. So from a Fregean standpoint, deflationism is simply not a viable position.

---

free variables introduced in the generalized substitution (which are unrestricted). It is easy to see that if a generalized substitution instance of B has a model, so does B itself, so the Conditional Possibility Principle is validated by the semantics of the appendix.

[13] This principal is valid as it stands on a free logic like that of Dana Scott's 'Existence and description in formal logic'. If one prefers a free logic like that of Tyler Burge's 'Truth and Singular terms', where each assertion of the form 'if $p(t_1, \ldots t_n)$ then $\exists x(x = t_1) \& \ldots \& \exists x(x = t_n)$' for atomic p is regarded as a truth of logic, then the principle must be modified slightly (say by redefining 'substitution instance' to mean a substitution instance in the normal sense conjoined with a clause of the form '$\exists x(x = t)$' for each term t in the sentence in which the substitutions are being made).

[14] Of course, knowledge of possibility is derivable from it: for example, that $\Diamond(AX_{NBG} \supset R')$, or that $\Diamond AX_{NBG} \supset \Diamond R'$; presumably, however, the Fregean view is that knowledge of possibility is never problematic if it is derivable from knowledge of necessity. (I count a knowledge claim as involving knowledge of possibility if its formulation contains at least one positive occurrence of '$\Diamond$' or at least one negative occurrence of '$\Box$'.)

I don't find this fact terribly upsetting, however, because I don't think that the Fregean viewpoint has a great deal of plausibility. In the first place, consider a point I made earlier, that part of our reason for believing that $\Diamond AX_{NBG}$ is the fact that we have been unable to derive any contradictions from AX_{NBG}. I argued then that this was a point that a deflationist could consistently recognize; I now want to observe that an advocate of *the Fregean position* could *not* recognize this point. For our inability to derive a contradiction from AX_{NBG} certainly doesn't give us reason to think that *actually* AX_{NBG}: if our reasons for believing that $\Diamond AX_{NBG}$ had to be based wholly on reasons for believing that actually AX_{NBG}, our inability to derive a contradiction from AX_{NBG} would be irrelevant to our knowledge that $\Diamond AX_{NBG}$.[15]

[15] It may seem that I am slurring over a complication. For it may seem that the fact that after persistent efforts we have not succeeded in deriving a contradiction from AX_{NBG} doesn't in itself provide evidence for $\Diamond AX_{NBG}$; rather, it provides evidence for a claim of *impossibility*, namely the impossibility of there being a derivation of a contradiction from AX_{NBG}. If this is right, then we need to establish a connection between this impossibility claim and the possibility claim $\Diamond AX_{NBG}$. Such a connection is established by the modal completeness theorem for first order logic, discussed in the next section.

How would this affect the argument in the text against the Fregean viewpoint? It might initially be thought to undercut the argument: for there is nothing in the Fregean viewpoint to rule out acceptance of the impossibility of deriving a contradiction from NBG on the basis of failures to find such derivations; and once that impossibility claim is accepted, it would appear that an application of the modal completeness theorem would yield knowledge that $\Diamond AX_{NBG}$. But of course the question is, how is the completeness theorem known? Knowledge of the modal completeness theorem is knowledge of possibility, and the proof of it sketched in section 4 assumes that $\Diamond AX_{NBG}$ (though doubtless we could make do with $\Diamond AX_M$ for a mathematical theory M somewhat weaker than NBG). Consequently, from a Fregean viewpoint it is hard to see how one could ever apply the theorem unless one already knew of an actual structure in which NBG (or the hypothetical weaker theory M) could be embedded. But that would mean that there would be no chance of adding to the credibility of the claim $\Diamond AX_{NBG}$ (or the claim $\Diamond AX_M$, if an appropriate weaker M is found) by persistently trying to derive a contradiction from NBG (or M) and consistently failing.

We see, then, that describing the epistemological situation as in the opening paragraph of this note would not help the Fregean. It would, though, create a problem for the non-Fregean (whether the non-Fregean be a platonist or a deflationist): for if our failures to derive a contradiction from AX_{NBG} only give reason to believe $\Diamond AX_{NBG}$ if we presuppose modal completeness, and that requires $\Diamond AX_{NBG}$, then such appeal to our failure to derive a contradiction is problematic from the non-Fregean position as well. Fortunately, however, the more complicated description of the epistemological situation seems wrong. It depends on thinking that the record of failures to find a contradiction could only enter into the epistemological picture as a premise to an enumerative induction. The right way to look at the matter, though, is as an inference to the best explanation: the assumption that $\Diamond AX_{NBG}$ is the most plausible explanation of the failure to find a contradiction in NBG. Modal completeness is irrelevant. (What is relevant is what I call in section 4 the 'weak modal soundness' of first order logic, but this is not something that one proves by $\Diamond AX_{NBG}$.)

In the second place, much of the motivation for the Fregean position is lost when we move from the crude formulation that Frege actually gives for his position (in the quotation above from §95 of *The Foundations of Arithmetic*) to the more defensible formulation that I have given. The motivation for the crude Fregean position is that it provides a simple solution to an epistemological problem, the problem of explaining the source of our knowledge of possibility. The crude Fregean position is that there really is no problem here: the source of our knowledge of possibility is just knowledge of actuality. The more defensible alteration of the Fregean position gives up this advantage: there is knowledge of possibility (not based solely on knowledge of necessity) that is not based on knowledge of actuality, but on 'reflection on relations of logical form'. The ability to 'reflect on relations of logical form' is supposed to allow us to know each instance of the schema '$\Diamond\psi \supset \Diamond\phi$' where ϕ is non-modal and ψ is a generalized substitution instance of it; but is it so clear that any motivated account of how we 'reflect on logical form' so as to come to this knowledge wouldn't also provide an account of how we know categorically some claims of form $\Diamond\phi$? After all, any claim of the same logical form as

(S) $\Diamond$(there are at least $10^{10^{10}}$ apples)

is also a logical truth. So why can't 'reflection on logical form' *show* that it is a logical truth? Why do we need to rest all our confidence in (S) on the claim that there actually are at least $10^{10^{10}}$ *somethings*? If we do need to rest knowledge of (S) on knowledge of actuality, that is rather surprising. What motivation is there for granting that we can have knowledge of possibility through 'reflection on logical form', but at the same time denying that knowledge of a simple possibility claim like (S) can be known by the same process?

Indeed, it seems to me that the Fregean position leads to quite counterintuitive consequences, for it seems clear to me that we have much more solid reason to believe (S) than to believe in the existence of $10^{10^{10}}$ entities of any kind. Certainly the claim that there are at least $10^{10^{10}}$ *physical* entities is not obvious. (I am inclined to believe it, for I am inclined to believe that regions of space are physical and that there are infinitely many of them; but I am much less confident of this than I am of (S).) And in my view, the claim that there are at least $10^{10^{10}}$ mathematical entities is far *less* obvious – in fact, I don't think there are *any* such entities – so I certainly wouldn't want to rest my belief in (S) on *that*. The idea that I should be as uncertain of (S) as I am of the claim that there are infinitely many physical or mathematical entities in the universe seems preposterous, and since the Fregean view has this consequence, I would need a much better motivation for that view before I could take it seriously.

The points I have made here for (S) arise for the claim '$\Diamond AX_{NBG}$' as well. That is, if this claim is true, so is every other claim of the same logical form; so why can't 'reflection on logical form' (whatever exactly that is) or whatever other process or combination of processes one needs to account for modal reasoning give us reason to believe it? Indeed, the example of '$\Diamond AX_{NBG}$' makes clear that the claim of even the crude Fregean view to be epistemologically pure was a hoax. The crude Fregean view was presented as having the epistemological advantage that it makes knowledge of possibility depend entirely on knowledge of actuality – but *mathematical* actuality is the only actuality that could work in the case of the claim '$\Diamond AX_{NBG}$', and once this is seen, it is hard to see how there is an epistemological advantage. The problem of how I know that it is *logically possible* that AX_{NBG} is 'solved' on the Fregean view by saying that I know that there *actually are* the entities that AX_{NBG} says there are and that they are interrelated as AX_{NBG} says. I find it hard to grasp how anyone could know (or have reason to believe) that such platonic entities actually exist (as opposed to being merely logically possible); or how anyone could know (or have reason to believe) that if such entities do actually exist, then they are related in one way rather than in some other. Consequently, the idea that I could explain my knowledge (or my reason to believe) that $\Diamond AX_{NBG}$ by reference to such knowledge of the platonic realm seems to me a total obfuscation of the real epistemological issues about knowledge of logical modalities.

3

Deflationism is, of course, a non-realist philosophy of mathematics: it holds that we cannot know (or have any reason to believe in) the existence of mathematical entities or the truth of ordinary mathematical claims; indeed, it would be natural to couple deflationism with the further claim that we have good reason to believe that there are no mathematical entities and hence that most ordinary mathematical claims are false.[16]

It has been widely held that the most serious difficulty facing any non-realist philosophy of mathematics is *the problem of application*: how can one account for the utility that reasoning about mathematical entities has for disciplines other than mathematics, if mathematics isn't construed in a realistic fashion?[17] Applied to deflationism, the problem

---

[16] Strictly speaking, the existentially quantified assertions will all be false and the universally quantified ones all vacuously true.

is: how can one explain the applicability of mathematics to disciplines other than mathematics, without assuming that ordinary mathematical claims (including those claims that assert the existence of mathematical entities) are true?

In a book I wrote several years ago I focused on one aspect of the problem of application: the problem of explaining the applicability of mathematics *to physical science* (and to everyday empirical reasoning), without assuming the truth of the mathematics that was being applied.[18] But there are other aspects to the problem of application, and a main task of the rest of this essay will be to say something about one of the more pressing aspects: the problem will be to explain the applicability of mathematics (in this case, proof theory and model theory) *to the study of logical reasoning*, without assuming the truth of the mathematics that is being applied.

Before turning to this main topic I want to say something about the applicability of mathematics to physical science and to ordinary empirical reasoning. Any account of the usefulness of a mathematical theory in dealing with the physical world will say that this usefulness depends on two things:

(a) the fact that the mathematical theory is 'mathematically good';
(b) the fact that the physical world is such as to make the mathematical theory particularly useful in describing it.

Different accounts of the usefulness of mathematics in application to the physical world will differ as to how (a) and (b) are to be elaborated.

A deflationist account of the application of mathematics must involve two claims. As regards (a), it must say that 'mathematical goodness' does not involve truth, but only something less demanding, such as consistency.[19] (This is strictly inaccurate, involving an inappropriate

---

[17] For instance, Frege says that 'it is applicability alone which elevates arithmetic from a game to the rank of a science' (Frege, *Grundgesetze der Arithmetik*, vol. ii, section 92.) The point has been most thoroughly developed by Hilary Putnam in *The Philosophy of Logic*.

[18] *Science without Numbers.*

[19] Of course, a deflationist can and will recognize that not all consistent mathematical theories (and not all mathematical theories that are *strongly* consistent in the sense shortly to be defined) are of equal mathematical interest – just as the platonist can and will say that not all *true* mathematical theories are of equal mathematical interest. What makes a mathematical theory interesting is a complicated matter – richness in consequences is one factor, relevance to prior work in mathematics and in science is another, elegance is a third and doubtless there are further factors still. There is no need to discuss such factors here, for they are ones that the platonist and the deflationist can agree on. What are important in the present context are the features of mathematical goodness that go beyond

cont. over page

semantic ascent. The more accurate formulation is that the deflationist must claim that in explaining the application of a mathematical theory, we do not need to assume the conjunction of its axioms, since that conjunction isn't logically true; he or she must claim that instead we need to assume only something weaker which *is* logically true, such as the result of prefixing the conjunction of the mathematical axioms with the modal operator '$\Diamond$'. For the moment I will forgo such accuracy for the sake of naturalness.) As regards (b), the deflationist must be able to formulate the facts about the physical world that make the mathematical theory 'fit it', without assuming in the formulation any standard mathematics (either the mathematical theory whose usefulness is being explained or some other mathematical theory). For if we have to assume the truth of standard mathematics anywhere in our account – in (b) or in (a) – then a deflationist would have to hold that such an account was unknowable.

So a deflationist has two tasks corresponding to (a) and (b) above. The task corresponding to (b) is by far the more difficult, but since I have treated it at length in *Science without Numbers* and since it involves issues rather far removed from those of the rest of this essay, I will say no more about it here. The task corresponding to (a) is much easier. I have discussed it too in my book and in a more elementary paper;[20] but here I will need to summarize quickly what I said about it.

My conclusion was that a mathematical theory needn't be true to be good; and, indeed, if it were true, this wouldn't be enough for it to be good, for a good mathematical theory must have a property that might be called *strong consistency* or *conservativeness* and that doesn't follow from truth alone. To say that a mathematical theory M is strongly consistent is to say roughly that if you take any theory T that says nothing about mathematical entities, and add T to M, then if T is consistent, so is T + M.[21] Although strong consistency doesn't follow

---

interestingness. For the platonist, a mathematical theory should not only be interesting, it should be *true*; and my question is, what serves the role for the deflationist that truth serves for the platonist?

[20] Essay 2. For a discussion of some more technical aspects of the issues surrounding (a) (including replies to some technical objections that have been raised), see essay 4.

[21] More formally, for any theory T let T* be the result of restricting all variables of T by the condition 'is not mathematical'. Restrict attention to theories T that don't contain the predicate 'mathematical' and also don't contain any specifically mathematical vocabulary such as 'set'. Then a mathematical theory M is strongly consistent if for any such T, if T is consistent then so is T* + M. (The restriction on the vocabulary of T is needed to rule out T implying that there are mathematical objects, or that there are objects such as sets that M implies are mathematical.) In cases of interest, the intended ontology of T will include no mathematical objects. Still, if M is an impure mathematical

cont. over page

from truth, it does follow from *necessary* truth; I believe that the widespread view of mathematics as necessarily true shows an implicit recognition of the importance of strong consistency. Strong consistency, however, is weaker than necessary truth, for strongly consistent theories needn't be true at all. As for ordinary consistency, this is not in general a sufficient requirement on a mathematical theory. There is an important class of mathematical theories ('pure' mathematical theories – those dealing with mathematical entities alone) for which ordinary consistency entails strong consistency; so for such theories the only requirement is that they be consistent in the ordinary sense. But there is another important class of mathematical theories (e.g., certain versions of set theory with urelements which play an important role in the application of pure mathematical theories) for which ordinary consistency does not entail strong consistency; and for these theories strong consistency is the important notion. I have argued in the works mentioned that in explaining the application of mathematics to the physical world we never need assume that the mathematics is true, we need only assume that it is strongly consistent (i.e., conservative).

This sounds like an account of application that is congenial to the deflationist: the truth of standard mathematics needn't be assumed in the account, so there seems to be no problem in reconciling the application of mathematics with the deflationist's claim that the truth of standard mathematics can't be known.

---

theory, the domains of T* and M will overlap, and so will their non-logical vocabularies. For instance, if M is the most useful version of set theory with urelements, the domains of M and of T* will include all non-mathematical urelements; and all the vocabulary of T will appear in M too, in the comprehension schemata. (Similarly, if M is the most useful version of impure number theory, it will contain an operator 'the number of x such that Fx', and through this the ontology and vocabulary of T will appear in the mathematics. For instance, one of the instances of the induction schema will be that if there is a planet with 0 moons, and if for each n such that there is a planet with n moons there is also a planet with n + 1 moons, then for all n there is a planet with n moons.) The fact that the ontology and vocabulary of T and M overlap explains why strong consistency doesn't reduce to ordinary consistency for typical impure mathematical theories.

The above definition of strong consistency is essentially the one given in *Science without Numbers*. (In the book I followed the standard artifice of regarding logic as ruling out the empty domain; because of this, I added some extra complexity in the definition of strong consistency, but there is no point in adding it here since the artifice has been dropped. I also misformulated the restriction on the vocabulary of T in fn. 8, though it should have been clear from the context that this was what was meant.) In the book I did not contemplate adding mathematics to theories that contained modal operators, for I was concerned only with adding them to physical theories and I did not (and do not) want to allow modal operators into nominalized physics. If we do consider the addition of mathematics to a theory T with modal operators, we might want to complicate the definition of T*; but there is no need to go into that here.

I think that this point is correct in spirit, but there is a difficulty that must be faced. For I have defined strong consistency in terms of ordinary consistency, and ordinary consistency is usually defined in terms of the existence of models. So won't the assertion that a theory is strongly consistent be an assertion about the existence of models? If so, then even though strong consistency doesn't entail truth, it is still hard to see how a deflationist could ever claim to know any theory to be strongly consistent. And if the deflationist can't claim to know that, it is certainly awkward for him or her to maintain that the strong consistency of a theory is essential to its application.

Many people have objected along these lines to the account of mathematics put forward in my book, and with considerable justification.[22] But from what I have said so far in this paper, it should be clear in outline how I now want to handle the objection. I want to say that in explaining the application of a mathematical theory M to the physical world, it is not strictly accurate to say that we need to assume the strong consistency of the mathematical theory. Rather, what we must assume is a certain modal claim, one which bears the same relation to the claim that M is strongly consistent that $\Diamond AX_M$ bears to the claim that M is consistent in the ordinary sense. The points I made several paragraphs back about the relation between strong consistency, ordinary consistency, truth and necessary truth should really have been made at the object level: instead of saying that strong consistency is entailed by necessary truth and entails consistency, but neither entails nor is entailed by truth, I should have said that the claim Q that is the modal analogue of the strong consistency of M is entailed by $\Box AX_M$, entails $\Diamond AX_M$ and neither entails nor is entailed by AX_M. Similarly, for the point about explaining the utility of M: instead of saying that we don't need to assume the truth of M, but only its strong consistency, I should have said that we don't need to assume AX_M but only Q.

That, it should be clear, is how I want to handle the objection. But *can* I handle it in this way? There is a technical difficulty to doing so, for there is a technical difficulty in figuring out exactly how Q (the modal analogue of strong consistency) is to be formulated.

Before turning to this technical difficulty, I want to discuss an earlier technical difficulty that I mentioned in section 1: the difficulty about

---

[22]   For example, Michael Resnik and David Malament in their reviews of my book; Charles Chihara in 'A simple type theory without platonic domains' and Michael Detlefsen in *Hilbert's Program: an essay on mathematical instrumentalism*. (For some reason Malament confines his objection to the case where a non-axiomatizable logic is at issue, but in fact it applies equally well to axiomatizable logics, since formal derivations are just as suspect from a nominalist viewpoint as models are.)

theories that are not finitely axiomatized. In section 1, I suggested that 'knowledge of the consistency of the theory of linear order' is really just modal knowledge: it is knowledge of the form $\Diamond B$, where B abbreviates the conjunction of the axioms of the theory of linear order. But suppose we consider a theory that is not finitely axiomatizable. Can't we know that such a theory is consistent? And how do we represent such knowledge as modal knowledge given our inability to conjoin all of the infinitely many axioms?

There is more than one way to respond to this objection; my current inclination is to respond by introducing a further logical device that will allow us to finitely axiomatize theories which, without the device, can't be so axiomatized. The device I have in mind is what is called a 'substitutional quantifier'. The name seems to me misleading: it is not a quantifier at all, as quantifiers are normally understood; rather, it is simply a device for representing sufficiently regular infinite conjunctions in a finite notation.[23]

The non-finitely axiomatized theories we ordinarily use are theories with quite regular infinite collections of axioms. For the theories consist of finitely many axioms plus finitely many axiom *schemata*: schemata like 'For every formula F, '$\exists z \forall y \, (y \epsilon z \equiv F)$' is to be an axiom.' To represent this finitely, we merely need to conjoin all the infinitely many sentences in the language of the form '$\exists z \forall y \, (y \epsilon z \equiv F)$'; and we can do that if we have a substitutional quantifier 'Π' with formulas as the substitution class, for we simply say '$\Pi F \exists z \forall y \, (y \epsilon z \equiv F)$.' (This is not a metalinguistic assertion but an infinitary conjunction of non-metalinguistic claims: it is no more metalinguistic than is a finite conjunction like '$\exists z \forall y \, (y \epsilon z \equiv y$ is a cat) & $\exists z \forall y \, (y \epsilon z \equiv y$ is a dog).') Anything that we normally regard as a mathematical theory can be finitely axiomatized using such a device, so any knowledge that we possess of the consistency of mathematical theories can be represented in the form $\Diamond B$ if we're allowed to use a substitutional quantifier in formulating B.

One advantage of this way of solving the technical difficulty about non-finitely axiomatized theories is that it solves the technical difficulty about the modal analogue of strong consistency as well. Strong consistency is defined in terms of ordinary consistency as follows:

(iii) for any theory T, if T is consistent, then so is T* + M;

here T* is the result of restricting all the quantifiers of T to non-mathematical entities. What is the modal analogue of this? It will differ

<hr>

[23] The view of substitutional quantification implicit in these remarks is set out in more detail in my review of Dale Gottlieb's book *Ontological Economy: substitutional quantification and mathematics*.

from (iii) in not being metalinguistic: instead of saying that if T is consistent, then so is T* + M, we will say that if $\Diamond AX_T$, then $\Diamond((AX_T)^*$ & AX_M). But then, how do we *generalize* such an object-level statement to all theories T? Again we must invoke a substitutional quantifier (or some other form of infinite conjunction); we must say

    (iii*) ΠB (if $\Diamond B$, then $\Diamond(B^*$ & AX_M))

My solution to the technical difficulty raised several pages back, then, is that 'knowledge of conservativeness' is really just modal knowledge of the form (iii*). Such modal knowledge does not require knowledge that AX_M (much less that $\Box AX_M$); and this is all to the good, since AX_M (and $\Box AX_M$) entails the existence of mathematical entities and consequently is not logically true. For certain mathematical theories, (iii*) will be stronger than the claim $\Diamond AX_M$, but even so, it is a logical truth. And the application of the theory M to the physical world requires only the logical truth (iii*), it does not require a claim like AX_M which asserts the existence of mathematical entities and hence is not logically true.

4

So far I've argued that for lots of purposes where we might seem to need notions like consequence and consistency and strong consistency or conservativeness, we really need only modal analogues of these notions, and that this is good because you can explain how facts involving the modal analogues are known without postulating knowledge of mathematical entities. In other words, in many contexts metalogical notions (at least notions of *semantic* metalogic) are dispensable in favour of corresponding object-level notions.

But an important problem remains for a deflationist: the problem of accounting for the utility of reasoning at the metalogical level rather than at the object level. The problem is especially striking for proof-theoretic reasoning, since here there seems to be no object-level analogue. An object-level assertion is one that makes no reference to sentences or formulas (or abstract analogues of them such as propositions); consequently, it can make no reference to axioms or rules of inference or formal derivations. It is hard to see how any such assertion could in any interesting sense be an analogue of an assertion that one sentence is (or is not) formally derivable from another, using a given system of logical axioms and logical inference rules.

How then are we to account for the utility of proof-theoretic reasoning? Traditionally, proof-theoretic concepts are defined in terms of mathematical entities, with the result that proof-theoretic reasoning becomes reasoning about mathematical entities. If we accept the usual definitions of proof-theoretic concepts, then a deflationist cannot regard proof theory as a subject of which we can have any knowledge. So how can a deflationist account for its utility?

There seem to be two ways for the deflationist to try to solve this problem. The first is to reject the usual definition of proof-theoretic concepts, and provide alternative definitions which make no reference to mathematical entities. The idea would be to show that if proof-theoretic notions such as formal derivability are understood in terms of these alternative definitions, then claims about formal derivability (etc.) can be known consistently with deflationism, that is, consistently with there being no knowledge of mathematical entities.

One way to try to work out this first approach would be to take derivability to be some sort of modal notion. First we could try to get a sufficiently powerful theory of actual inscriptions, without introducing modality: in terms of such a theory, we could explain notions like 'e is a well-formed inscription', 'e and f are type-identical inscriptions', 'd is (an inscription that constitutes) a derivation (according to system F)', and various predicates of inscriptions that describe them structurally ('being an A-inscription', where A is an expression type). I believe that this part of the project could be worked out in first order logic (though some care is needed because there are no means here to make recursive definitions explicit). The second part of the project would be to make some modal extension: in it, we might hope, we could understand 'A is derivable' to mean 'it is possible that there is a derivation whose last line is an A-inscription', instead of (as the platonist would have it) as meaning that there actually exists a certain type of abstract sequence of abstract analogues of the symbols. This has considerable initial attraction as an account of the ordinary meaning of 'derivable'. I am, though, reluctant to introduce a new type of possibility beyond strictly logical possibility here, unless we can define it from strictly logical possibility plus other acceptable notions; and there are substantial difficulties that must be overcome for the project of doing this.[24] We could avoid these

---

[24] For instance, there are at least two obstacles to taking the relevant sort of possibility to be consistency with the first order theory that one obtained in the first part of the project. The first obstacle arises from the fact that logical possibility is thoroughly anti-essentialist: this poses a problem for translating sentences in which 'derivable' occurs in the scope of a quantifier. ('He uttered an underivable inscription' would always be false on the naive translation.) The natural way to solve this problem would be to take all

cont. over page

difficulties with additional logical devices like a substitutional quantifier, but these might make the appeal to modality unnecessary anyway. I will not pursue these matters here.

In any case, carrying out this programme wouldn't really solve the more general problem raised two paragraphs back. The problem was for the deflationist to account for the utility of proof theory without assuming the truth of mathematics. And in presenting the problem it was assumed that proof theory meant *platonistic* proof theory. The first deflationist approach to this problem says that there is a nominalistic proof theory that is just as good as platonistic proof theory, and that the nominalist has no difficulty in accounting for the utility of *that*. But unless more is said, it looks as if this is merely changing the subject from the original question, which was how the utility of the platonistic theory is to be explained.

My primary interest then will be with the second deflationist approach to the problem of explaining the utility of proof theory: the approach of trying to explain the utility of platonistic proof theory without assuming it true. I will explain shortly how this second approach can be carried out.

But first I want to shift attention from proof theory back to semantics. The problem with which I began this section – the problem of accounting for the utility of mathematics in metalogical reasoning – is a problem that arises for semantics as well as for proof theory (though it may initially seem less striking a problem in the case of semantics, since there we have object-level analogues of our metalogical notions). And again, there are two different approaches that one might be inclined to take.

The first approach would be to reject the usual model-theoretic definitions of semantic consequence and similar notions, and propose alternative definitions instead. Can this be done? In a sense it can. We can say

(5) B is a semantic consequence of Γ (where Γ is a finite list of sentences together with a finite list of schemata) if and only if □ (if all members of Γ are true, then B is true).

'quantification in' to be substitutional. The second obstacle is that the consistency with axiomatic proof theory (even axiomatic platonistic proof theory) of the existence of a proof is not sufficient for provability in the normal sense: incompleteness theorems give cases of unprovable formulas where the assertion that there is a proof is consistent with proof theory. The natural way to solve this would be to say that for A to be provable, the existence of a proof of A must be compatible with a (nominalistic or platonistic) proof theory stated in a powerful logic that can rule out derivations that are not genuinely finite: say a logic with the quantifier 'there are only finitely many', or a logic with a substitutional quantifier or other device of infinite conjunction.

This definition would be objectionable if the word 'true' here were used in a 'transcendent' sense, that is, in a sense in which 'Snow is white' wouldn't have been true if 'white' had meant 'green'; for in that transcendent sense the modal claim on the right of (5) is going to be false even if B is a consequence of Γ in the usual sense. But let us understand 'true' instead in the 'immanent' sense,[25] that is, as applicable most directly to one's own language and as obeying there the principle that $\square$('Snow is white' is true if and only if snow is white). We can define such an immanent sense of 'true' using substitutional quantifiers: S is true if and only if $\Pi p(S = \text{'p'} \supset p)$. Or alternatively, one can follow Grover, Camp and Belnap and regard 'true' (or at least 'true' in the immanent sense) not as a predicate at all but (roughly speaking) as simply the means by which substitutional quantifiers with sentences as substituends are represented in English.[26] In either case, it turns out that (5) is equivalent within standard mathematics to $\square$(if AX_Γ then B), where AX_Γ is a conjunction of all the axioms in Γ (using substitutional quantifiers to conjoin the instances of the schemata). On this alternative to the usual way of defining consequence, what I earlier called the modal analogue of a claim that one thing was a consequence of another wouldn't really be an *analogue* at all, it would simply be what the consequence claim *means*.

I don't attach a great deal of philosophical significance to the possibility of defining semantic consequence in this nonstandard way. My reason for mentioning it is only to point out that even if it is adopted, it leaves an important problem unsolved: namely, how is a deflationist going to explain the utility of *model-theoretic* definitions of semantic consequence? Even assuming that it is somehow better to define consequence modally in the way just indicated (a claim on which I take no stand), still model-theoretic semantics has proved enormously useful; and it is not immediately evident *why* it should be useful if consequence is really to be defined modally via an immanent notion of truth, or if consequence claims are to be rejected as strictly speaking unknowable and only modal analogues of them are to be claimed as knowable. So a deflationist must give an account of why standard uses of model theory are legitimate even if model theory isn't true, just as he or she must give an account of why standard uses of platonistic proof theory are legitimate even if platonistic proof theory isn't true.

In order to provide such accounts, we must first ask to what uses model-theoretic semantics and proof theory are standardly put; only

---

[25] This use of the terms 'transcendent' and 'immanent' was suggested by (but isn't quite the same as) Quine's use of these terms in *Philosophy of Logic*.

[26] 'A prosentential theory of truth'.

then can we ask whether these uses are explainable from a deflationist viewpoint. I will not attempt an exhaustive account of the uses to which model-theoretic semantics and proof theory are put, but it seems to me that the central uses are as devices for finding out about logical possibility. Model theory is used for this purpose via the instances of the following two schemata: the *model-theoretic possibility schema*

(MTP) If there is a model for 'A' then $\Diamond A$;

and the *model existence schema*

(ME) If there is no model for 'A' then $-\Diamond A$.[27]

And proof theory is used for this purpose via the *modal soundness schema*

(MS) If there is a proof of '-A' in F then $-\Diamond A$,

which holds for any reasonable formal system F for any fragment of logic; and via the *modal completeness schema*

(MC) If there is no proof of '-A' in F then $\Diamond A$,

which holds for certain areas of logic (i.e., certain types of sentence A) and certain sufficiently strong formal systems for those areas of logic.

From the platonist standpoint, all of the instances of these four schemata are true (for the appropriate formal systems F, in the case of the last two schemata).[28]

What *justification* might a platonist offer for these schemata? I'll focus on MTP and MS, since these seem the evidentially primary ones. (The

[27] If A contains free variables, interpret 'model' in (MTP) and (ME) to mean a model together with an assignment function for the free variables.

[28] Actually there is some question as to whether we should expect (ME) to hold for arbitrary logics. Even for first order logic, it seems somewhat surprising that (ME) holds (in the same way that the 'Skolem paradox' seems somewhat surprising): just as the universe of classes is too big to form a countable model, it is too big to form a class and hence too big to form any model at all; so just as it seems somewhat surprising that the sentence AX_{NBG} that formulates the theory of this universe of classes should have a countable model, it seems a bit surprising that it should have a model at all. But the classical completeness theorem shows that it does have a model if it is not formally inconsistent; this makes (ME) a set-theoretic consequence of (MS) in the case of first order logic, which surely makes it platonistically acceptable in the case of first order logic. Still, it may not be acceptable for arbitrary logics. It seems clear that if it fails for some logic – for example, if there is a sentence A formulatable in that logic which expresses enough about the 'intended model' of NBG to preclude there being any class that can be a model of A – then the usual model-theoretic definition of consistency should be regarded as extensionally incorrect for that logic: the imagined sentence A is intuitively consistent, even if it has no model.

other two follow from these, of course, by the classical completeness theorem.) In the case of MTP, each first order instance (instance where the instantiating formula is non-modal) is almost immediate from an instance of the Conditional Possibility Principle of section 2 (together with NBG and an instance of 'A $\supset$ $\Diamond$A').[29] (I'll discuss the case where the instantiating formula is modal later.) In the case of MS, though, the situation is more delicate. To simplify the discussion, we'll pick an F in which in any application of a rule of inference, the formula being inferred is a logical consequence of the formula from which it is inferred.[30] A platonist might be tempted to argue for the validity of MS (as applied to F) by induction. More accurately, we could argue informally that for all sentences B, if there is a proof of B in F then $\ulcorner\Box B\urcorner$ is true, by induction on the length of the proof: certainly $\ulcorner\Box B\urcorner$ is true when B is a logical axiom of any reasonable F, and if B is directly inferable from $B_1, \ldots, B_n$ then $\ulcorner\Box(B_1 \& \ldots \& B_n \supset B)\urcorner$ is true, and so using the induction hypothesis $\ulcorner\Box B\urcorner$ is true. In particular then, for all sentences A, if there is a proof of $\ulcorner\text{-}A\urcorner$ in F then $\ulcorner\text{-}\Diamond A\urcorner$ is true; by disquotational properties of truth the instances of (MS) follow.

But this informal induction goes beyond the modal consequences of standard mathematics, if 'standard mathematics' is taken in the normal way: for in the induction I have utilized a notion of truth (for sentences in the modal language) that has not been defined. Indeed, besides the usual problem with defining truth, and hence carrying out the induction, that the Tarski indefinability theorem poses, there is a further problem: what would the recursion clause for formulas that begin with a logical possibility operator be? In the appendix, in defining truth *in a model*, I use the clause

'$\Diamond$A' is true *in model M* $\equiv$ there is a model in which A is true.

[29] For any formula B containing predicates $p_1, \ldots, p_k$, let M, $E_1, \ldots, E_k$ be variables not in B, and let $\hat{B}$ be the generalized substitution instance that results from B by restricting quantifiers and free variables by the predicate 'x is in M' and by replacing (for each i and each $v_1, \ldots, v_{n_i}$) '$p_i (v_1, \ldots, v_{n_i})$' by '$<v_1, \ldots, v_{n_i}>$ is in E_i'. Also, let B^+ be the existential closure of $\hat{B}$. (Example: if B is '$\forall y (p_1(y,z))$', $\hat{B}$ is '$z\epsilon M \& (\forall y\epsilon M)(<y,z>\epsilon E_1)$', and B^+ is '$\exists M\exists E_1\exists z[z\epsilon M \& (\forall y\epsilon M)(<y,z>\epsilon E_1)]$'.) $\Diamond\hat{B}$ is equivalent to $\Diamond B^+$; this plus the Conditional Possibility Principle yields $\Diamond B^+ \supset \Diamond B$. So to get the instance of MTP we need only argue in NBG for 'If there is a model M and an assignment function s in which B comes out true, then $\Diamond B^+$.' But in fact we can inductively argue in NBG that if there is such an M and s then *actually* B^+; so since possibility follows from actuality, we have the instance of MTP.

[30] In the case of non-modal quantification theory, the system of Hunter's *Metalogic* has this property: instead of containing a rule of universal generalization it contains the principle that the universal generalizations of quantificational axioms are quantificational axioms. The analogous strategy can be used in modal systems: avoid a necessitation rule by taking necessitations of all axioms to be axioms.

(For perspicuity I consider only the case where A is a sentence.) It might be natural to say by analogy

(*) '$\Diamond$A' is true $\equiv$ there is a model in which A is true.

But if 'standard mathematics' is taken as a first order axiomatic theory, this won't do very well as part of a recursive definition of truth, for then '$\Diamond$A' together with standard mathematics won't in general entail '"$\Diamond$A" is true.'[31] Reason: by Gödel's second incompleteness theorem, there are models of axiomatic set theory NBG in which the sentence 'NBG has a model' will be false. In such a model, '$\Diamond$NBG' will be true (it's true in all models), but on the proposed definition of truth '"$\Diamond$NBG" is true' will be false in the model.

I am not denying, of course, that a platonist should accept the material equivalence (*) (for sentences A not containing terms like 'true' – see footnote 31). Indeed, assuming that '"A" is true' is equivalent to 'A' (for sentences not containing 'true'), (*) is simply the conjunction of MTP and ME, which I have said a platonist ought to accept. The argument is, though, that there are models of set theory in which the left to right direction of (*) fails, so that *barring independent support of MS and hence of ME*, (*) does not have the kind of necessity we would like in a recursion clause of a definition of truth.

We could avoid this difficulty about the '$\Diamond$A' clause by weakening MS to a *non-modal* soundness schema, i.e. by dropping the operator '$\Diamond$' in it, and at the same time restricting its instances to non-modal formulas. But even this wouldn't help with the basic problem: arguing for even this restricted non-modal soundness inductively requires a truth predicate, and we don't have one by the Tarski indefinability theorem.

Indeed, the Gödelian example shows not only that there are problems in formalizing the intuitive proof of MS modally within standard mathematics: it shows that MS is not a consequence of standard mathematics alone, even for non-modal A and even in the strong modal logic given in the appendix. Let CON_{NBG} be the statement that there is no F-proof of not-NBG (where F is a proof procedure for first order logic). Then an instance of MS – indeed, an instance with a purely first order sentence as the substituend for the schematic letter 'A' – is

If -CON_{NBG} then -$\Diamond$NBG;

equivalently,

(**) If $\Diamond$NBG then CON_{NBG}.

---

[31] The sentences A involved here don't contain 'true' or related terms, so the lessons of the semantic paradoxes do not do anything to make this conclusion palatable.

By Gödel's second incompleteness theorem, CON_{NBG} is not derivable from NBG. So there is a first order model in which NBG holds and CON_{NBG} doesn't. But the model theory in the appendix takes first order models to be modal models as well, so NBG doesn't imply CON_{NBG} modally either. But NBG does imply $\Diamond$NBG modally, so (**) cannot be a modal consequence of NBG.[32]

At this point in the original version of this essay I introduced a nonstandard sense of 'standard mathematics', employing substitutional quantifiers and an ω-rule governing them, from which MS does follow modally by an inductive proof like the one recently sketched (and in which (*) becomes a reasonable recursion clause for truth). It now seems to me, though, that this made the presentation of several points confusing, and the weight put on substitutional quantification may seem suspicious. It seems simpler just to say that a platonist just accepts MS, even though it is not provable in standard mathematics from more elementary modal principles. It is worth remarking (though how important this is I'm not sure) that in the modal logic of the appendix, the non-modal instances of MS are enough to generate (in standard mathematics in the usual sense) all instances of MTP and even of ME, including those where modal sentences are the substituends.[33]

I have been discussing the epistemological status, from a platonist perspective, of MTP and MS and the other lettered schemata. It is clear that once these schemata are available, the platonist can use ordinary

---

[32] In fact, of course, even the weakening of MS that drops the possibility operator is not a consequence of standard mathematics.

[33] For both MTP and ME, one uses an induction on the modal degree of the formula A that is the substituend (generalizing MTP and ME to formulas as in fn. 27). If A is of modal degree $n + 1$, let A^* be the result of taking each sub-formula of form $\Diamond B$ for which B is degree 0 (i.e., non-modal) and replacing it by 'B v -B' if B has a model and by 'B & -B' otherwise. Now we've proved the degree 0 instances of MTP, and the degree 0 instances of ME follow by completeness from the degree 0 instances of MS, which we're assuming. Using these, we get (in S5) that if B non-modal then $\Box\{(\Diamond B) \equiv (\Diamond B)^*\}$, and by a subinduction on the depth of the embedding of $\Diamond B$ in A, $\Box\{A \equiv A^*\}$. Consequently, (1) $\Diamond A \equiv \Diamond(A^*)$. Also, another trivial subinduction shows that any model is a model of A if and only if it is a model of A^*, and consequently (2) A has a model if and only if A^* has a model. But (1) and (2) and MTP for A^* (which is of modal degree n) yield MTP for A; and analogously for ME.

Incidentally, the reliance on the degree 0 instances of MS is essential, even for MTP: for instance, if $\Diamond$NBG but NBG has no model, then $-\Diamond$NBG has a model (indeed, it is true in *any* model), but $-\Diamond-\Diamond$NBG by the S5 axiom, so we have a degree 1 violation of MTP. Still, a given instance of MTP, with A the substituend formula, can be proved from NBG plus n degree 0 instances of MS, where n is the number of occurrences of logical operators in A. It is easy to calculate what the required instances of MS are (inspection of the proof above shows how to do it), and in a typical case most of them will be provable in NBG anyway so they won't really need to be added.

platonist model theory and proof theory for finding out about possibility and impossibility. But how is this of any help to a deflationist, who denies that the existence of mathematical entities and the truth of mathematical theories can be known? Assume that we construe 'proof' and 'model' in the usual platonist way. Then if there are no mathematical entities, ME and MC are simply invalid: their antecedents are true irrespective of whether $\Diamond$A or -$\Diamond$A. And if there are no mathematical entities, MTP and MS are only *vacuously* true: they are useless as an aid to finding out about possibility and impossibility, because their antecedents can never be fulfilled. Admittedly, one could do a bit better by considering *concrete* proofs (made by actual physical inscriptions) and *concrete* models. By construing 'model' concretely, we'd get a 'weak MTP' that is non-vacuous (it would be little more than a restatement of the Conditional Possibility Principle, of course); but it would generate the logical possibility of rich structures only from controversial empirical premises. The 'weak MS' obtained by construing 'proof' concretely would be even more severely limited, since rigorous proofs in formal systems are rarely given for anything that is in the least complicated. Apparently, then, the deflationist has a problem.

In fact, though, the problem is easily solved. Instead of MTP and MS (or perhaps, in addition to their weak versions), the deflationist employs the following modal surrogates:

(MTP$^{\#}$) If $\Box$(NBG $\supset$ there is a model for 'A') then $\Diamond$A

and

(MS$^{\#}$) If $\Box$(NBG $\supset$ there is a proof of '-A' in F) then -$\Diamond$A.

From these and the classical completeness theorem, one can derive

(ME$^{\#}$) If $\Box$(NBG $\supset$ there is no model for 'A') then -$\Diamond$A

and

(MC$^{\#}$) If $\Box$ (NBG $\supset$ there is no proof of '-A' in F) then $\Diamond$A,

in the case where first order sentences are the only substituends; indeed, ME$^{\#}$ can be argued to hold also where modal sentences are substituends.[34] The deflationist can use the hatched schemata in pretty much the same way the platonist used the unhatched ones: to find out that A is, or is not, logically consistent, it suffices to derive a model-theoretic or proof-theoretic statement from standard mathematics.

Could it be claimed that the deflationist has less reason to believe MTP$^{\#}$ and MS$^{\#}$ than the platonist has to believe MTP and MS? I do

---

[34] This will be clear from the remarks on MTP$^{\#}$ below, in conjunction with fn. 33.

not see how this could be made plausible. First let's consider the instances of MTP$^\#$ in which the instantiating formula is non-modal. Earlier (footnote 29) I presented a platonist argument for the corresponding instances of MTP, within standard mathematics; in effect, then, I showed that for non-modal A

$\Box$(If NBG & there is a model for 'A', then $\Diamond$A.)

From this it follows (in S4) that

(MTP*) If $\Diamond$(NBG & there is a model for 'A'), then $\Diamond$A.

From this and the assumption $\Diamond$NBG, one gets the instance of MTP$^\#$. So the platonist's argument from NBG to MTP yields an argument from $\Diamond$NBG to MTP$^\#$. The deflationist does of course need to assume $\Diamond$NBG in order to accept MTP$^\#$, but if my earlier arguments in section 2 are correct, such consistency claims are ones that a deflationist can have perfectly good reason to believe. In any case, the deflationist's epistemological burden is strictly weaker than that of the platonist, who must believe not only that $\Diamond$NBG but that *actually* NBG.

The above reasoning can be extended to instances of MTP$^\#$ where the instantiating formula is modal. In footnote 33 I presented a platonist argument for an arbitrary instance of MTP: the argument was from NBG plus a few instances of MS (which ones being easily calculable from the substituend sentence in MTP). Letting NBG$_A$ be NBG with these added instances, the reasoning of the previous paragraph shows that the platonist's argument from NBG$_A$ to the instance of MTP yields a deflationist argument from $\Diamond$NBG$_A$ to MTP$^\#$. (Actually to a slight strengthening of MTP$^\#$, one obtained by replacing 'NBG' in it by 'NBG$_A$'.) Again, the claim that the deflationist needs, $\Diamond$NBG$_A$, is a logical consequence of the one that the platonist needs, and I don't see how it can be plausibly argued that the deflationist is in worse epistemological shape than the platonist here.

What about the epistemological status of MS$^\#$? In the case of MS, the reader will recall, the platonist had no hope of rigorously deriving it from NBG (even in the strong modal logic of the appendix); nonetheless, MS is a claim that a platonist ought to accept as a primitive modal assumption. The deflationist, similarly, can accept MS$^\#$ as a primitive modal assumption. Alternatively, the deflationist can derive each instance of MS$^\#$ from *the possibility of* what the platonist assumes: i.e., from $\Diamond$(NBG + the corresponding instance of MS). (This is a trivial derivation in S5.) Either way, the deflationist seems to be in pretty good epistemological shape.

A platonist might respond to this by saying that in the case of MS we have an informal (and unformalizable) inductive argument in its favour; but that the deflationist has no such inductive argument in favour of sentences of form $\Diamond(\text{NBG} + \text{B})$ where B is an instance of MS, and so must ride piggyback on a platonist argument that he or she does not accept. I think, though, that this is wrong. To say that one accepts an informal inductive argument that cannot be formalized in one's theory is to say in effect that one accepts a stronger theory. In the case of the argument for MS, it might plausibly be argued that we are implicitly employing ordinary set theory in some modal framework, to which a truth predicate has been added. (Without the modal framework, the addition of a truth predicate suffices to prove *non*-modal soundness.)[34a] If the modal framework is chosen appropriately – and the truth predicate is a truth predicate for modal sentences as well as non-modal ones – modal soundness should be derivable analogously. But now, if the platonist can appeal to such a powerful theory S, the deflationist can appeal to $\Diamond\text{AX}_\text{S}$.[35] And since the instances of MS follow from S, the instances of MS$^\#$ follow from $\Diamond\text{AX}_\text{S}$, by the same argument as for MTP$^\#$.

Another alternative for the platonist who wants a proof of MS is to employ some sort of ω-rule. That is, even without a notion of truth we can give a general method for proving each instance of

(MS$^\text{k}$)   If there is a proof of '-A' of length k, then -$\Diamond$A

(where k is any numeral and A any formula): exhaustively search the derivations of length k, and the result will either falsify the antecedent or yield a proof of the consequent. But without a truth predicate, we cannot in ordinary set theory get from these instances to the generalization that for all k MS$^\text{k}$ holds. But if an ω-rule were adopted, a platonist could get this generalization from the various MS$^\text{k}$, or from the argument for their provability. My response of course is that if a platonist is allowed to employ an ω-rule, so is the deflationist. But if so, then the same sort of argument as above gets us from the platonist's 'derivation' of MS in the expanded logic to a deflationist's 'derivation' of MS$^\#$ in the expanded logic. However the platonist twists and turns in an effort to avoid taking MS as simply a primitive assumption, the deflationist can twist and turn too.

I conclude that the deflationist has no more difficulty in using platonistic model theory and proof theory in finding out about possibility and impossibility than does the platonist.

In section 3, I distinguished two 'problems of application': the problem of application of mathematics to the physical world, and the

---

[34a] Here I assume that it is a truth predicate for sentences of set theory that don't contain it, that it is allowed to occur in the separation and replacement schemas, and that 'proof' in MS is understood as 'proof not employing the new truth predicate'.

[35] S won't be finitely axiomatized; but we can avoid the use of substitutional quantifiers here, by using the theory consisting of the possibilization of each finite conjunction, as discussed in the postscript.

problem of application of mathematics to the study of logical reasoning. I have just outlined a solution to the latter problem; but readers may be puzzled by a disanalogy between this solution and the solution offered in *Science without Numbers* to the problem of application to the physical world. In solving both problems of application, I tried to legitimize a certain kind of instrumentalism about mathematics: I tried to argue that platonistic physics and platonistic metalogic were usable even if not true. But in *explaining* why the usability of these theories didn't depend on their truth, I had to do more work in the case of platonistic physics than in the case of platonistic metalogic. That is, the explanation given in my book for the legitimate usability of platonistic physics turns on the existence of a nominalistic physics. But the explanation I have just given for the legitimate usability of platonistic proof theory doesn't require the existence of a nominalistic proof theory. (I have stated that such a nominalistic proof theory may be possible, but my explanation of the usability of platonistic proof theory in finding out about possibility and impossibility did not turn on this.) What accounts for the difference?

The answer is that physics has an explanatory function: you need physical theories to explain physical phenomena. According to the form of nominalism I accept, one should not junk a platonistic explanation of a phenomenon unless there is a satisfactory nominalistic explanation to take its place. It is because of this principle that a satisfactory nominalistic formulation of physical theories is required. Now, the main role of platonistic proof theory is not explanatory. If I use proof theory as an aid to discovering whether B follows from A, it is not because the proof-theoretic principles are in any way explanatory of the fact that $\Box(A \supset B)$ or $-\Box(A \supset B)$; the proof theory is solely an instrument of discovery, and needn't be replaced by some other theory about which we must take a non-instrumentalist attitude.

I have not said that proof theory has *no* explanatory function, but only that its *central* function is not explanatory. There is a sense in which proof theory can be used to explain. Suppose I want to explain the historical fact that no one has ever produced a physical inscription which constitutes a formal derivation, in Kleene's system of logic, of an explicit contradiction. Intuitively, the reason that no one has ever produced such an inscription is that it is impossible that there be one; and here I think 'impossible' can be taken to mean 'logically inconsistent with certain assumptions we make about physical inscriptions'. Any codification of those assumptions is a theory of physical inscriptions, and from it we can explain the historical fact in question. One way to codify those assumptions about physical inscriptions is to formulate platonist proof theory and then add a principle saying that there is a certain kind of homomorphism mapping physical inscriptions into expressions in the platonist sense. This codification produces a platonist explanation of the historical fact in question. I also think that assumptions about physical inscriptions which are adequate to the purposes at hand can be stated nominalistically (without using devices going beyond first order logic, in this case); if so, then we will have a nominalistic explanation of the historical fact. I take it, though, that the use of proof theory to explain such historical facts is of less importance than its use in finding out about possibility and impossibility; and in the latter use, the platonist proof theory does not serve an explanatory function, and so no nominalistic proof theory is required.

I have been discussing a disanalogy between my treatments of physics and of proof theory; how does model theory enter into the picture? That is, in explaining the legitimate usability of platonistic model theory, did we (as with platonistic physics) need to develop a nominalistic analogue of the platonistic theory? Or (as with the central applications of platonistic proof theory) did we not? This question is largely verbal: it depends on whether we regard modal logic itself as an analogue of platonistic model theory. If we do so regard it, then model theory is like physics, and the earlier sections of this paper were devoted largely to developing the nominalistic analogue as a necessary prelude to explaining the legitimate usability of platonistic model theory.[36] If we do not so regard it, then model theory is like proof theory in its most central applications: we did not need a nominalistic analogue of model theory because model theory doesn't serve to explain anything, but simply serves as a tool for enabling us to find out more easily about possibility and impossibility. The difference between these two viewpoints is merely a difference between ways of looking at what has been done: whatever the viewpoint, the argument earlier in this section shows how one can explain the legitimate usability of platonistic model theory without assuming its truth.

5

In this essay I have been advocating the view that all mathematical knowledge that isn't straightforwardly empirical is knowledge of a purely logical sort. By 'mathematical knowledge' here I do not mean knowledge of the claims of mathematics. According to the view I have advocated (deflationism) there *is* no mathematical knowledge in that sense. Rather, by 'mathematical knowledge' I mean the sort of knowledge that those

---

[36] The prelude is necessary, for as I argued in essay 2, there is no way to explain the legitimate usability of metalogic by conservativeness alone if the underlying logic is taken to be non-modal.

who know a lot of mathematics have a lot of and those who know little mathematics have little of. If deflationism is false, this will include knowledge of mathematical claims; but whether deflationism is true or false, it will include a lot of knowledge that isn't knowledge of mathematical claims.

As hinted several times, some of the knowledge that separates those who know lots of mathematics from those who know only a little is straightforwardly empirical. A set theorist typically knows which axioms of set theory are generally accepted within the mathematical community, which theorems have been proved, which unsolved problems are generally regarded as important; and any algebraist knows that mathematicians have thoroughly developed a generalization of vector space theory in which the role that fields play in vector space theory is played by unitary rings. All these sorts of knowledge are empirical knowledge about the mathematical community. Besides such knowledge about the mathematical community, mathematicians typically have other empirical knowledge that non-mathematicians tend to lack; for example, typically there will be various complicated empirical claims about physical space which they know because they know them to follow from the empirical fact that physical space can be locally approximated as a Cartesian power of the real numbers. If one were to attempt a realistic account of all of the knowledge differences that separate a typical mathematician from a typical non-mathematician, I think that such differences in empirical knowledge would play a large role. So (contrary to what the title of this paper might suggest) a great deal of mathematical knowledge is the sort of straightforward empirical knowledge that no one could possibly regard as logical. The interesting question, however, concerns the mathematical knowledge that remains when this straightforward empirical knowledge is ignored. The deflationist claim that I have defended is that the only such knowledge there is is purely logical – even on a conception of logic according to which logic can make no existence claims. (It would not seem to me to be terribly interesting to say only that such knowledge was logical on a broader conception of logic like that of the logicists – a conception on which logic guarantees the existence of a realm of platonic entities. See footnote 3.)

The deflationist view is reminiscent of, but importantly different from, a position that has been called 'deductivism' or 'if-thenism.' Deductivism is usually characterized as the view that when someone asserts a typical mathematical statement (e.g., that there are infinitely many primes), what he or she really means is that this statement follows from a certain body of other mathematical statements.[37] (*Which* body of other

---

[37] See, for instance, ch. 3 of Michael Resnik, *Frege and the Philosophy of Mathematics*.

mathematical statements? The standard axioms of the field of mathematics in question, if the field has an accepted axiomatization; otherwise, some other body of claims implicit from the context. Deductivists tend to be a little vague about this, often entirely ignoring the situation where there is no generally accepted axiomatization.)

One of the major differences between this and deflationism is that deflationism, unlike deductivism, does not claim that mathematical claims mean anything other than what they appear to mean. Instead of saying that mathematical assertions don't mean what they appear to mean, the deflationist says that what they literally mean can't be known: the knowledge that underlies a mathematician's assertions is not what those assertions literally say. I'm afraid that many readers will still find this implausible, but it certainly seems to me less implausible than the claims about meaning made by the deductivist. (The implausibility of the deductivist position is especially evident in the case of a mathematical assertion A made in the absence of a generally accepted axiomatization. The deductivist must select some one body of other mathematical statements, and claim that what is meant in saying A is really that A follows from this other body of statements. But the deflationist, since he or she makes no claim about meaning, need not single out any one body of other mathematical statements as relevant: *no* bodies of other statements are relevant to what the assertion of A *means*, and *lots* of bodies of assertions are relevant to what the mathematician who asserted A *knows*, since a great many distinct pieces of knowledge of the interrelation of A with other mathematical claims may have been part of the motivation for asserting A.)

Indeed, the deflationist can easily handle a problem that is often thought to sink deductivism. Consider a mathematician asserting a claim which he or she knows not to follow from previously accepted axioms; he or she intends it as a new axiom. What does he or she mean when asserting this mathematical claim? If one takes deductivism entirely literally, then according to the deductivist, the mathematician must mean either

(a)   that the new axiom follows from the old axioms

or

(b)   that the new axiom follows from the system that consists of the old axioms plus the new axiom.

But the mathematician doesn't believe (a), and (b) is totally trivial. Since the whole point of the deductivist's claims about what the mathematician means is to make what he or she means directly reflect part of the knowledge that led to the assertion, both of these alternatives are intolerable.

For a deflationist, on the other hand, the situation where the mathematician introduces a new axiom poses no special problem. The reason is that the deflationist does not accept the programme of trying to represent the knowledge that leads the mathematician to make the assertion in the meaning of the assertion. The kind of knowledge that typically leads a mathematician to assert a new axiom is clear enough: it is knowledge that the axiom (in conjunction with previously accepted axioms) has certain desirable consequences and doesn't seem to have undesirable ones. In other words, it is knowledge of the logical interrelations of the proposed axiom with other mathematical claims, which is just the sort of knowledge that the deflationist wants to allow anyway. So the situation where someone asserts a mathematical claim because of the consequences he or she knows it to have is no more of a problem for a deflationist than the situation where he or she asserts it because he or she knows it to be a consequence of previously accepted claims. In general, I think it would be extremely surprising if careful attention to the sociology of mathematical practice turned up features of that practice that couldn't plausibly be handled along deflationist lines.

6

It is often supposed that one of the things that differentiates those who know a lot of mathematics from those who know only a little is that the former have considerable knowledge of a realm of platonic entities such as sets, numbers and tensors – entities that bear neither causal nor spatio-temporal relations to us or to anything we can observe. If this were correct, there would be a considerable problem in explaining how knowledge of such a realm could be attained. The strategy of this paper has been to point out various facts that aren't about such a platonic realm, facts which mathematicians typically know and non-mathematicians typically don't. These facts include empirical facts, like the facts mentioned early in section 5; and they include logical facts, like the facts about logical possibility that I have stressed in the bulk of the paper. I see no reason to believe that there is a further kind of fact – non-empirical and non-logical – that the mathematician also knows. Therefore, I see no reason to suppose that the mathematician has knowledge of the existence of mathematical entities or the truth of ordinary mathematical claims.

Appendix

In this appendix I will sketch an alternative to Kripke's model theory for modal logic, one which will give the intuitively correct results about which sentences involving the 'logically possible' operator are logically true. The model theory will be, like Kripke's, platonistic, for it will presuppose a large body of pure set theory. What a deflationist should say about the status of a platonistic model theory such as this is discussed in section 4 of the essay. (The purpose of the model theory is not to confer intelligibility on the modal operator. In my view, logic stands on its own; it doesn't need model theory for its intelligibility. Indeed, it is hard to see how a logic could get its intelligibility from the model theory for it, since one would need the logic in understanding (e.g., in being able to reason from) the model-theoretic assertions.)

In the model theory for first order logic, we say that a sentence is *logically true* if and only if it is *true in all models*. Here, a *model* consists of a set of objects (the entities that *exist in the model*) plus a stipulation as to which things if any the predicates are *true of* in the model, which things if any the names *denote* in the model and so forth. We also need the notion of an *assignment function for a model*: it is a partial function that assigns entities that exist in the model to all, some or none of the variables of the language.[38] Given a model M and an assignment function

---

[38] It is standard in first order logic to restrict consideration to models in which at least one object exists and in which all names denote. When this restriction is made, assignment functions are taken to be total functions, that is, they assign something to every variable. I have tacitly lifted this restriction in the body of the appendix, since it would be more glaringly anomalous in modal logic than it is in first order logic. (There are two ways of lifting the restriction that seem about equally reasonable, those of Scott and of Burge – see fn. 13. Strictly speaking, the definition of 'model' given in the appendix is applicable only to Burge's system, but nothing of substance would be altered if we complicated the definition slightly so as to apply to Scott's. In particular, the definition of 'model' for a modal logic based on Scott's free logic would be the same as for the non-modal Scott free logic.)

s for M, it is possible to recursively define what it is for a formula of the language to be *true in M relative to s*. We then define a formula to be *true in M* if and only if it is true in M relative to s, for *every* assignment function s for M; and we define a formula to be *logically true* if and only if it is true in M for every model M.

How should we generalize this to modal logic? Kripke's approach is to keep the idea that logical truth is truth in all models, but to redefine the notion of model: in the case of the system S5 (which is the one of interest for present purposes), a model is to be a non-empty set of possible worlds, one of which is designated as actual; each possible world is determined by a set of objects that exist in that possible world, plus a stipulation as to what things in that world the predicates are true of in that world in that model, plus similar stipulations for names and other primitive vocabulary of the language. A sentence of the form '◇A' will be true in a model just in case A is true in at least one possible world in the model. For '◇A' to be logically true, it must be true in all models, *and hence in particular it must be true in all models in which there is only one possible world* (i.e. in which there are no possible worlds other than the actual world of the model). It is clear that there is no way that this can happen unless A itself is true in all models, that is, unless A itself is logically true. That is the curious feature of Kripke's definition of logical truth for modal logic that I noted at the beginning of section 2.

I propose an alternative way of generalizing the definition of logical truth for sentences of first order logic to a definition appropriate to modal sentences. As on Kripke's approach, we are to retain the idea that logical truth is truth in all models. In addition, *we retain the definition of model used in first order logic*. (We do not introduce possible worlds; rather, a model will be in effect just the 'actual world portion' of a Kripke model.) The *only* difference between the proposed definition of logical truth for modal logic and the usual definition of logical truth for first order logic is that in recursively defining truth in a model we need an extra clause that will handle formulas beginning with '◇'.

Moreover, the rule for '◇' will be a lot like the rule for '∃' used in first order logic. In first order logic the rule for '∃' is as follows:

> '∃xB' is true in M relative to s if and only if B is true in M relative to s*, for some s* that assigns something to the variable x and that is just like s except in what it assigns to the variable x.

Note that in this rule we quantify over assignment functions, leaving the model fixed. I propose that in our rule for '◇' we quantify over models and assignment functions together:

'◇B' is true in M relative to s if and only if B is true in M* relative to s*, for some model M* and some assignment function s* for M*.

If B is a sentence (i.e., has no free variables), all reference to assignment functions can be proved irrelevant. That is, in that case the rule reduces to

'◇B' is true in M if and only if B is true in M* for some model M*.

It is clear that on this approach, unlike Kripke's, such sentences as

$$\Diamond \exists x \exists y (x \neq y)$$

and

$$\Diamond \exists x (x \text{ is an electron})$$

will come out true in all models, and hence logically true.

The approach that I have sketched for defining logical truth for modal sentences has its roots in chapter 5 of Carnap's *Meaning and Necessity*;[39] though much of what Quine found abhorrent about Carnap's ideas has been dropped. In the first place, the approach I have sketched does not rely in any way on the idea of meaning. As I remarked in the text, the basic modal logic is formulated in such a way that it does not reflect 'meaning relations among predicates' if such a notion be recognized. (Though if one does recognize such a notion, a derivative modal logic can be obtained which does reflect such relations.) In the second place, the treatment of free variables that I have given does not require the introduction of 'individual concepts', and it is thoroughly anti-essentialist in that no formula of the form '◇B' is true in a model with respect to one assignment function unless it is true in that model with respect to every other assignment function. (Again, however, the notion of logical

---

[39] An account even more similar than Carnap's to that given here is that of Nino Cocchiarella, 'On the primary and secondary semantics of logical necessity'. But Cocchiarella's method of dealing with variables and their interaction with modal operators seems to me unacceptable: for instance, it leads to Ramsey's bizarre conclusion that 'It is possible that there are at least $10^{10^{10}}$ objects' is logically false if the world happens to contain less than $10^{10^{10}}$ objects, but is logically true if the world happens to contain at least $10^{10^{10}}$ objects. (Cf. the last section of Ramsey's paper 'The foundations of mathematics'.) Despite this difference between Cocchiarella's account and the one I have offered, most of Cocchiarella's philosophical remarks about logical truth in modal logic apply to my account as well as to his own. (Some of the others who have advocated something in this general ballpark are Richard Montague (*Formal Philosophy*, ch. 1); Jaako Hintikka ('Standard vs. nonstandard logic') and Dana Scott ('On engendering an illusion of understanding'). Charles Parsons discusses a similar viewpoint in 'Quine on the philosophy of mathematics'.)

possibility can be used to introduce derivative notions of possibility which are essentialist in various ways.) A third respect in which my views modify Carnap's (though in this case, not a modification that Quine would favour) is that Carnap's idea was to regard modal concepts as derivative from semantic concepts. On my view it is, if anything, the other way around, as long as it is purely logical possibility that is in question.

Postscript

1 Motivation. The paper does not sufficiently emphasize that the idea of modal metalogic is appealing independently of anti-platonism. For more on this, see section 5 of the introduction to this volume.

2 Substitutional Quantifiers. In section 3 of this essay I used 'substitutional quantifiers' (viewed as devices of infinitary conjunction) for two purposes. But they weren't really needed.[1]

Mathematical and physical theories are standardly axiomatized with finitely many axioms and in addition finitely many axiom schemata each of which has infinitely many instances. The use of such axiom schemata is usually thought (rightly or wrongly) to make sense independently of any devices (such as substitutional quantifiers) which would enable us to encapsulate the schemata into single axioms. The idea is that even without reducing such schemata to single axioms, we can explain what it is to accept the schema: to accept the schema is simply to accept each of its instances. Let us assume here that this usual attitude is correct.[2]

The first use to which I put substitutional quantifiers in this essay was to express modally the idea that a given infinitely axiomatized theory in mathematics or physics was consistent. What is primarily of interest here (I would argue) is mathematical and physical theories expressed in a logic (such as first order logic) for which compactness holds. In this case, the consistency of the whole theory

[1] In the original version I made another use of substitutional quantifiers in section 4, but I have dropped that discussion from the current version, as explained in the text.

[2] There are two possible grounds for questioning it. One is that the idea of accepting infinitely many things makes no sense, unless there is a finite bunch of principles that we accept from which they obviously follow. Another is that even if we do suppose that it makes sense, still accepting the instances is too weak an account of what it is to accept the schema: accepting the schema is like accepting the conjunction of all the instances, and we could accept each instance without accepting the infinitary conjunction. I sympathize with both these grounds, and as a result think that the use of substitutional quantifiers as a device of infinitary conjunction is ultimately needed in metalogic. But my point here will be that the need has nothing to do with the issues about modality and consequence *per se*.

is the same as the consistency of each of its finite conjunctions; so we can explain what it is to believe that T is consistent by saying that we believe each sentence $\Diamond(T_1 \& \ldots \& T_n)$, where T_i s are axioms. The 'infinitariness' involved in accepting the consistency of T seems no more problematic than the 'infinitariness' of accepting T itself. If we have an infinitely axiomatized theory that we believe is logically true, then expressing our belief is also unproblematic: we believe each claim of form $\Box T_i$. (These beliefs of course entail $\Box(T_1 \& \ldots \& T_n)$, for each finite conjunction).

If we want to deny the consistency of an infinitely axiomatized theory, though, and if we don't know where the inconsistency lies, then we have a bit of a problem. The inconsistency of T is of course the same as the logical truth of the denial of T; but we may not know how to express the denial of T without conjoining the sentences of T, and consequently we may not be able to deny the consistency of T in this case. Here substitutional quantifiers (or some more or less equivalent device of infinite conjunction, such as a disquotational truth predicate – see section 4 of essay 7) are needed. Note, though, that the reason they are needed doesn't have anything to do with the modal representation of consistency *per se*, it has to do rather with the representation of *negation* for infinitely axiomatized theories. We would have the same problem if we wanted to express the belief that an infinitely axiomatized theory was false if we didn't know which part of it was false.

At any rate, in this essay the issue of denying the consistency of an infinitely axiomatizable theory never arises.

The second use to which I put substitutional quantifiers was in formulating the strong consistency or conservativeness of a mathematical theory. But again we don't really need substitutional quantifiers: to accept the conservativeness of M is simply to accept each instance of the schema

$$(C) \text{ If } \Diamond B \text{ then } \Diamond(B^* \& M_1 \& \ldots \& M_n),$$

where B is any sentence, B^* is the result of restricting B to non-mathematical entities, and $M_1, \ldots, M_n$ are axioms of M. To be sure, this only directly expresses the conservativeness of M with respect to *finitely axiomatized* nominalistic theories B;[3] but the conservativeness with respect to infinitely axiomatized nominalistic theories follows. Proof: suppose T is infinitely axiomatized and consistent, i.e., $\Diamond(T_1 \& \ldots \& T_m)$ whenever $T_1, \ldots, T_m$ are any of its axioms. If M is conservative with respect to finitely axiomatized theories, then (since the starring operator distributes over conjunction) $\Diamond(T_1^* \& \ldots \& T_m^* \& M_1 \& \ldots \& M_n)$ for any $T_1, \ldots, T_m, M_1, \ldots, M_n$. But this is just what it should mean to say that $T^* \& M$ is consistent.

The upshot of this is that while I believe in the use of substitutional quantification for certain purposes (viz., the purposes that others would use disquotational truth for), they aren't really needed in the context of this essay, and it would have been tactically advantageous simply to formulate conservativeness by schema (C).

---

[3]   (C) holds even for non-nominalistic B: the effect of the * operator is to reinterpret such a B in an unintended way as nominalistic.

3   Modal soundness and conservativeness. I should have pointed out that my modal soundness principle MS is really the platonistic formulation of another kind of conservativeness claim, and that MS$^{\#}$ is the deflationist's version of the same claim. What MS and MS$^{\#}$ amount to is that typical applications of proof theory don't yield conclusions you wouldn't get otherwise, just as what I've been calling conservativeness says that other typical applications of mathematics don't yield conclusions you wouldn't get otherwise. It is illuminating to write the deflationist's version of both kinds of conservativeness together, in a common format. MS$^{\#}$ generalizes from NBG to mathematical theories more generally as the schema

$\quad$ (MS$^{\#}$) If $\Box$(M$_1$ & . . . & M$_n$ $\supset$ there is a proof of '-A' in F), then $-\Diamond$A,

while (C) contraposed yields

$\quad$ (C) If $\Box$(M$_1$ & . . . & M$_n$ $\supset$ -A*), then $-\Diamond$A.

4   Strengthening MTP$^{\#}$ and MC$^{\#}$; the modal version of the Kreisel squeezing argument; and modal analogues of provability. The modal derivation of MTP$^{\#}$ (from the Conditional Possibility Principle) proceeds by the derivation of a stronger claim:

$\quad$ (MTP*) $\Diamond$(NBG & there is a model of 'A') $\supset$ $\Diamond$A.

(It's stronger given $\Diamond$NBG, anyway.) From this and classical completeness, we get a strengthened modal completeness principle:

$\quad$ (MC*) $\Diamond$(NBG & there is no F-derivation of '-A') $\supset$ $\Diamond$A,

where F is a typical formal system for quantificational logic. Analogous strengthenings of MS and ME, by contrast, are false, as footnote 24 shows. (In the original version of this essay I asserted an analogue of ME* *when NBG was replaced by a strengthened mathematics that included an ω-rule*; but here I am dropping that strengthening.)

In this essay I do not make anything of the fact that we can derive MTP* rather than merely MTP$^{\#}$, but it is important. Consider the Kreisel analysis (introduction, section 5) of the philosophical significance of the completeness theorem for first order logic. Kreisel argues in effect that by taking 'it is logically consistent that' as a primitive operator (here symbolized as '$\Diamond$') governed by the principles MTP and MS, the role of the completeness theorem is to enable us to prove the biconditionals

$\quad$ $\Diamond$A $\equiv$ there is as model of 'A'

and

$\quad$ $\Diamond$A $\equiv$ there is no F-derivation of '-A'.

We *prove* these, without regarding either as *defining* consistency. This is of course a platonistic analysis, since an anti-platonist cannot accept those biconditionals, but it would be nice if the anti-platonist could say something analogous. The analogue will of course involve using MS# instead of MS; and we get an especially clean analogue if we use MTP* (rather than MTP#) instead of MTP. The analogue goes like this: MS# gives as a *necessary* condition for ◇A that ◇(NBG & there is no F-derivation of '-A'); MTP* gives as a *sufficient* condition for ◇A that ◇(NBG & there is a model of 'A'). But the completeness theorem in the form the anti-platonist accepts it – that is, □{NBG ⊃ (there is no F-derivation of '-A' ⊃ there is a model of A)} – shows that the necessary condition entails the sufficient condition: there is no room between them, so each is both necessary and sufficient. That is,

(i) ◇A ≡ ◇(NBG & there is a model of 'A')

and

(ii) ◇A ≡ ◇(NBG & there is no F-derivation of '-A').

Or equivalently,

(iii) □A ≡ □(NBG ⊃ 'A' holds in all models)

and

(iv) □A ≡ □(NBG ⊃ there is an F-derivation of 'A').

From a platonist point of view, it should be no surprise that these hold. Consider (iv): the platonist analysis using MTP and MS yields

□A ≡ there is an F-derivation of 'A';

but a platonist will also accept

there is an F-derivation of A ≡ □(NBG ⊃ there is an F-derivation of 'A'),

since 'there is an F-derivation of x' is a Σ_1 formula, and since NBG is ω-consistent. (1-consistency is actually all that's needed.) The two indented conditions give us (iv). Of course, the derivation of (iv) in the previous paragraph was independent of this platonist derivation of it: it relied only on the Conditional Possibility Principle (to get MTP*) and MS#. (The use of MS# is the anti-platonist analogue of the assumption of 1-consistency.)

Despite the fact that the operators ◇ and □ in (i)–(iv) occur on the right hand side of the biconditionals as well as on the left, I think that (i)–(iv) should allay any doubts about the clarity of those operators. (At least, they should allay doubts about the clarity of the operators as applied to non-modal sentences, which is where I primarily want to apply them.)

Incidentally, since 'there is an S-derivation of x' is Σ_1 for any formal system S (not just for formal systems of quantification theory), then from a platonist point of view we have in general that □(NBG ⊃ there is an S-derivation of '-A') if and only if there is an S-derivation of 'A'. (The same holds if 'NBG' is weakened to 'platonistic proof theory'.) In footnote 24 I raised two difficulties for a modal definition of S-derivability. The kind of definition I was considering there was

◇(Proof theory & there is an S-derivation of 'A'),

and the second difficulty was that (because of Gödel's theorem) this has to be extensionally inadequate if the proof theory is formulated in first order logic alone. We now see that in a sense we could have solved the difficulty by a different kind of modal 'definition' of S-derivability, namely

$\Box$(Proof theory $\supset$ there is an S-derivation of 'A').

It should be noted, though, that this has no plausibility whatever as an account of the ordinary notion of derivability: it is a modal *surrogate* of derivability, not a modal *analysis*. If one wants a modal analysis, one still has to introduce richer logical structure to go with the modality, as discussed in footnote 24.

5 On the strong modal logic employed. In the appendix to this essay I sketch a platonist model theory for a quite strong modal logic: it consists in tacking onto S5 *all truths of form '$\Diamond A$', where A is non-modal.* (More exactly, one tacks all such truths onto a certain anti-essentialist version of quantified S5.) However, no positive use is made of anything but an axiomatized fragment of this.

I do need a strong possibility axiom, such as $\Diamond$(NBG). (It might be better to try to find more obvious possibility principles from which this could be shown to follow, but I have not felt the need to pursue this: $\Diamond$(NBG) seems obvious enough.) I also need two other principles: the Conditional Possibility Principle of section 2 and, ultimately, MS$^{\#}$. (These are consequences of the strong logic with all true possibility statements in it, but they need separate inclusion with an axiomatized fragment.)

I did make a negative use of the strong logic with all true possibility statements in it: I argued that the modal soundness claim employed by the platonist, namely MS, isn't a modal consequence of standard mathematics even in that strong logic. But of course here the only role of the strong logic is to serve as an upper bound on the logic that might reasonably be employed.[4]

It is worth pointing out that to define logical truth in such a way that it obeys a strong modal logic does not commit one to any strong claims about logical *knowability*. And in this essay I make a sharp distinction between logical truth and logical knowability: indeed, most of section 2 is devoted to considering the view that even if claims like $\Diamond$(NBG) are logically true, they cannot be logically known. Of course, I considered the view only to reject it. But my argument against the view was certainly *not* an argument that the distinction between logical truth and logical knowability collapses.

---

[4] The only other role that I gave to the strong logic was in my extension of MTP and ME from non-modal formulas to modal formulas: it is of course hardly surprising that one would need the strong modal logic there, since MTP and ME, as applied to modal formulas, are modal principles about truth of modal formulas in a model, and the definition of model and of truth in a model for modal formulas were designed with the strong modal logic in mind. Presumably if one wants to isolate a weaker modal logic one could find a notion of model and of truth in a model that is appropriate to that weaker logic, and given such a construal of truth in a model, MTP (and perhaps ME) could be shown in the weaker logic to hold for modal formulas as well as for non-modal ones. But I see little reason to pursue this: indeed, I'm not sure that MTP and ME are of much importance in their application to modal formulas.

6 Logical Knowledge. My claim that mathematical knowledge is just logical knowledge (insofar as it isn't just knowledge about the mathematical community or knowledge-how instead of knowledge-that and so forth) may suggest to some the idea that mathematical knowledge is indefeasible. This would be false (as the discovery of the inconsistency of Cantorian set theory illustrates dramatically), and was not intended: for I take logical knowledge to be defeasible too. In particular, the paper argues that claims of consistency are to be construed as logical, and they are clearly defeasible. (In the essay, I attempted to remove any suggestion of indefeasibility, by sometimes shifting from claims of what we can logically *know* to claims of what we can have logical *reason to believe*.) Also, the claim that knowledge or rational belief is logical was not meant to preclude its having a somewhat inductive character: indeed, I argued that knowledge of consistency of certain theories is at least partly based on the idea that if the theories were inconsistent we would probably have discovered it by now, but that this didn't prevent the knowledge from being logical. What I primarily meant to be saying, in calling knowledge of the consistency of mathematics logical, was that this knowledge did not have to be based on knowledge of that mathematics itself (or some stronger mathematics; or some theory about other entities as epistemologically problematic as mathematical entities, such as possible worlds).

4

On conservativeness and incompleteness

In 'Conservativeness and incompleteness' Stewart Shapiro develops in detail a Gödelian construction which is alluded to sketchily in my book *Science without Numbers* and argues that this construction has implications that undercut an important part of the philosophical project of that book.[*] I don't agree with this assessment and will explain why in this essay. I shall also be replying to some closely related objections which I have heard or read but which have not reached published form (including some objections which have been widely circulated by Saul Kripke); and I will be explaining some things that are not explained very well in the book.

1

A central claim of *Science without Numbers* is that mathematics does not need to be true to be good. The property that it needs instead of truth I called 'conservativeness', which I defined roughly as follows:

> A mathematical theory S is *conservative* if, for any nominalistic assertion A and any body of such assertions N, A is not a consequence of N+S unless A is a consequence of N alone.

A *nominalistic assertion* is an assertion whose variables are all explicitly restricted to non-mathematical entities. When S is a pure mathematical theory, i.e., a theory whose variables range over mathematical entities alone, then the conservativeness of S is an obvious consequence of its consistency. But impure mathematical theories, such as set theories

---

[*] Shapiro 1983. I am grateful to Shapiro for sending me his paper before publication and for a fruitful correspondence about the issues it raises. I would also like to thank John Burgess and Penelope Maddy for helpful discussions of some issues related to the ones that Shapiro raises, and to thank Maddy again for useful comments on an earlier draft.

with urelements, have variables ranging over mathematical and non-mathematical entities alike; and for some such set theories conservativeness is a bit stronger than consistency. Nevertheless, I argued that if we were to discover of a set theory that it was not conservative, this would be' overwhelming grounds for rejecting it, and that we have virtually as good reason to think that standard set theory is conservative as we have to think that it is consistent.

Shapiro points out that one might take conservativeness in either of two senses, semantic or proof-theoretic, depending on whether one understood the term 'consequence' in the usual sense (semantic) or understood it in terms of derivability. Since we will shortly be contemplating logics that, unlike first order logic, admit no complete proof procedure, the distinction is crucial; and in footnote 30 of *Science without Numbers* I explicitly opted for the (more natural) semantic construal of the definition. (In that footnote I point out that, earlier in the book, before the apparent need to go beyond first order logic had been mentioned, I had implicitly invoked the completeness of first order logic to pass back and forth between the official definition of conservativeness (which I understood semantically) and a proof theoretic analogue of it. But, I go on to say, 'in the case of second order logic there can be no such completeness theorem: here, we must stick to semantic notions throughout.' And I go on to show that we can give platonistic arguments for the *semantic* conservativeness of set theory in the context of second order logic, but that these arguments do not extend to *proof theoretic* conservativeness.)

But Shapiro thinks that this choice of semantic conservativeness over proof theoretic conservativeness was philosophically the wrong choice. He gives two reasons for this, which I will consider in a moment; he then proves that if I *had* intended the notion of conservativeness to be understood in the proof theoretic sense, my claims of conservativeness would have been incorrect for logics with no complete proof procedures. Shapiro grants that in the semantic sense mathematical theories are conservative; thus in the early part of his paper (i.e., before the final eight paragraphs) his objection is not the technical objection it might at first appear to be, but rather the philosophical objection that I focused on the wrong notion of conservativeness.

What are Shapiro's arguments for thinking that I made the wrong choice? His first argument is that some of what I say in explaining the utility of mathematics presupposes that it is the proof-theoretic notion that is at issue: I say that the utility of mathematics is that it enables us to *provide shorter deductions* of things that without the mathematics could still be *deduced* but only in a lengthier fashion. Shapiro is right to complain that such remarks (and there are a lot of them) don't sit well with the claim that *semantic* consequence is what is at issue. I wish I had put things differently. What I should have said is that mathematics is useful because it is often easier to see that a nominalistic claim follows from a nominalistic theory plus mathematics than to see that it follows from the nominalistic theory alone. (Indeed, this alteration in what I said seems preferable independently of the issues at hand.)

Shapiro's second reason for thinking that it is a proof-theoretic rather than a semantic notion of consequence that I should be interested in is deeper: the claim is that, on a nominalistic position like that defended in my book, one is going to have a hard time understanding a notion of logical consequence that goes beyond anything explainable in proof-theoretic terms. A full-scale discussion of this objection would involve consideration of a complicated set of issues which are rather far removed from those in the rest of this essay. But, to put it roughly and baldly, my reply is that the relevant notion of consequence can be explained modally (it is not possible for the premises to be true and the conclusion false) and that the modality can be understood without explanation in terms of platonistic entities (possible worlds, models, etc.). For a more accurate statement of the position and some argument in its support, I refer the reader to a recent paper of mine 'Is mathematical knowledge just logical knowledge?'[1] (The position is one that I developed since I wrote my book; unfortunately there is no discussion at all in the book about how metalogical relations are to be understood.) Meanwhile, I will simply observe that to the extent that there is a difficulty for the nominalist in explaining semantic consequence, there would seem to be an equal difficulty in explaining proof-theoretic consequence, i.e., derivability: after all, the standard (non-modal) explanation of this is in terms of the existence of abstract sequences of abstract expression-types, no token of which need ever have been written or spoken. If one insists on a non-modal explanation of consequence and hence of conservativeness, the proof-theoretic notion seems as problematic to a nominalist as the semantic; so I don't see that one can motivate using a proof-theoretic rather than a semantic notion of conservativeness in this way.

These remarks are not intended to deny that there is something worrying about the use of logics with no complete proof procedures: it does strike me as at least somewhat worrying (though as far as I can see, no more for the nominalist than for anyone else). I have been concerned, however, only with the conditional question: given that we are to use a logic where semantic consequence and proof-theoretic

[1] Essay 3 of this volume.

consequence diverge, is there any good reason why a defender of nominalism should focus on proof-theoretic conservativeness rather than semantic conservativeness? The answer to this, I think, is no. (There will be a bit more discussion of this claim at the end of section 4.)

Whatever the philosophical merits of this view, it is clear that Shapiro does not disagree with the conservativeness claim I made in *Science without Numbers*, namely that mathematical theories are semantically conservative. I have been surprised to find that a number of people seem to think that even this claim is false; I think, however, that their arguments are clearly based on a confusion. Baldly put, their arguments work by pointing out that there are theories to which the addition of mathematics does not provide a semantically conservative extension. In every case, however, the theories to which the mathematics is added already contain reference to mathematical entities: they are not nominalist theories; so the claim that mathematical theories are semantically conservative is in no way undercut.[2]

2

As I have said, a main goal of *Science without Numbers* was to argue that mathematical theories needn't be true to be good; the property they need instead of truth is (semantic) conservativeness. To say this is not to say that all semantically conservative theories are in all respects equally good; a nominalist need no more hold this than a platonist need hold that all true mathematical theories are in all respects equally good. A platonist and a nominalist can both admit that in certain respects the theory of real numbers is 'better' (e.g., more interesting, richer in mathematical consequences and more important for physics) than the theory of ordered fields that are closed under square root but are not real-closed, despite the fact that a platonist holds both to be true and a nominalist (as well as a platonist) holds both to be conservative.

An important line of objection to the nominalist claim that mathematics needn't be true but need only be conservative is the Quine-Putnam argument that you need to assume the truth of mathematics in order to do science. In my opinion, the only acceptable way to counteract this argument is to show that nominalistic resources are adequate to the statement of good scientific theories. This claim is not a consequence of the conservativeness of mathematics: the conservativeness of mathematics tells you what happens when you add mathematics to nominalistic

---

[2] The most popular examples of the fallacy I am alluding to concern the addition of mathematics to theories that contain the platonistic claim (C_p) formulated at the beginning of section 4.

theories; it doesn't say anything about the availability of sufficiently interesting nominalistic theories.

One way one might try to argue that nominalistic resources are adequate to the statement of good scientific theories would be to argue for the following claim:

> (E) For each platonistic scientific theory, there is a nominalistic theory of which the platonistic theory is a conservative extension.

But if no restrictions are made as to what sets of sentences that are closed under consequence count as theories, (E) is trivial: clearly each platonistic theory T is a conservative extension of the 'theory' consisting of the nominalistic consequences of T. (Even if we require that the platonistic and nominalistic theories be recursively axiomatized, (E) is still vacuous in the case where the underlying logic is first order: take the nominalistic 'theory' to consist of the Craigian reaxiomatization of the nominalistic consequences of T.) In order to reply to the Quine-Putnam argument along the lines of (E), one would first have to strengthen (E) so as to rule out the 'uninteresting' nominalistic theories; and it is not at all clear how this can be done in any precise way. For this reason, among others, it was not a goal of my book to argue for anything along the lines of (E). It is possible that some readers (not Shapiro) have become confused on this point because they have confused (E) with the claim that mathematics is conservative. My previous paragraph should make clear that this *is* a confusion: after all, (E) or any strengthening thereof is an assertion about the existence of a sufficiently wide variety of nominalistic theories, and this is something that the assertion that mathematical theories are conservative does not claim.

Let us use the phrase 'conservative sub-theory claim' to refer to any claim concerning the existence of nominalistic sub-theories of which a given platonistic theory is a conservative extension. Then there is a conservative sub-theory claim in my book. Unlike (E), it is a claim about one specific theory (Newtonian gravitational theory), though, as I remark there, it extends to many other physical theories of interest. (Just how far it extends isn't clear). Moreover, the claim holds for Newtonian gravitational theory only if that theory is formulated within a rather strong logic. In the last chapter of the book I noted that, if we restrict ourselves to standard first order logic, we do not get an interesting conservative sub-theory result. (Of course, we still get uninteresting results of the sorts noted in the previous paragraph.) But, I tried to argue, this does not seriously undermine the ability to respond to the Quine-Putnam argument. It is this last point that Shapiro is challenging in his final eight paragraphs. Again, I do not think the objection is the technical objection that Shapiro makes it seem — indeed, the technical result that Shapiro relies on here is one I discussed in my book. Rather, the objection is philosophical. But in order to state the philosophical issues clearly, it will be worth reviewing some of the technical background. Such a review will also be relevant to some of the issues that were discussed in section 1.

3

The basic strategy of *Science without Numbers* was to take physical theories stated in terms of numerical functors and to try to restate them in terms of comparative predicates instead. That is, instead of using such functors as 'the gravitational potential of x' (understood as having a numerical value), we are to use such predicates as 'the difference in gravitational potential between x and y is less than that between z and w.' These comparative predicates are to be taken as primitive, rather than (what my English rendering of them would suggest) as defined in terms of the numerical functors. The possibility of assigning a numerical gravitational potential $\psi(x)$ to each object x in such a way that the usual numerical laws of physics hold is to be regarded as a *derivative* fact; for it is to be explained in terms of 'intrinsic' laws, i.e., laws stated without reference to numbers, in terms of comparative predicates only.

In carrying out this idea I relied heavily on a body of pre-existing theorems called *representation theorems*. These theorems show how various numerical functors can be 'generated out of' underlying comparative predicates if those predicates are axiomatized in a certain way. What the pre-existing representation theorems didn't do is show how numerical laws stated in terms of the various numerical functors can be restated in terms of the various comparative predicates that 'generate' those functors. The main technical work in my book (chapter 8) was devoted to showing how to state numerical laws (e.g., differential equations such as Poisson's equation for the gravitational field) using comparative predicates only. The result of all this was an *extended-representation theorem*: this says that, for any model of a certain theory N that uses various comparative predicates but no numerical functors,

> (1) there are
>> (a) a 1–1 spatio-temporal co-ordinate function ϕ (unique up to generalized Galilean transformation) mapping the space-time of the model onto the quadruples of real numbers,
>> (b) a mass density function ρ (unique up to a positive multiplicative transformation) mapping the space-time of the model onto an interval of non-negative reals, and

    (c) a gravitational potential function ψ (unique up to positive linear transformation) mapping the space-time onto an interval of reals,

all of which 'preserve structure' (in the sense that the comparative relations defined in terms of these functions precisely coincide with the comparative relations used in N); moreover, the laws of Newtonian gravitational theory in their functorial form hold if ϕ, ρ, and ψ are taken as the denotations of the relevant functors.

Now, it is immediately evident that if this extended-representation theorem, as just stated, is to hold, N cannot be formulated in first order logic. For the extended-representation theorem as I have stated it clearly implies that any model of N is uncountable, and the downward Löwenheim-Skolem theorem tells us that that can't be true for (consistent) first order theories. (Similarly, the compactness theorem, or the upward Löwenheim-Skolem theorem, tells us that any first order space-time theory according to which space-time is infinite will have models in which the set of space-time points has a cardinality larger than the real numbers, and such a conclusion is also incompatible with the stated representation theorem holding.) In fact, the logic that I used to formulate N is quite a strong logic which I called 'the complete logic of the *part of* relation' or 'the complete logic of Goodmanian (i.e., mereological) sums';[3] it is sort of a nominalistic analogue of the second order logic that has unary predicate variables as its only higher order variables. But within this logic, the extended-representation theorem holds. The extended-representation theorem implies that N plus standard mathematics entails standard platonistic Newtonian gravitational theory. Moreover, N is a nominalistic theory, and so it follows from the conservativeness of mathematics that N (even without the mathematics) has all the nominalistic consequences that standard platonistic gravitational theory has.

In fact, if we take the mathematics involved in standard platonistic Newtonian gravitational theory to be a recursively axiomatized first order mathematical theory, then N will have nominalistic consequences that standard platonistic gravitational theory does not have. If on the other hand, we take the mathematics involved in standard platonistic gravitational theory to be second order set theory of the sort formulated by Richard Montague,[4] then N will have precisely the same nominalistic

---

[3] The theory I called 'N' in the book also used the quantifier 'there are only finitely many'. But, as I noted in the book (p. 93 and n. 71), this quantifier can be dropped in the presence of the logic of the *part of* relation; and for present purposes I imagine this done.

[4] 'Set theory and higher order logic'.

consequences as platonistic gravitational theory has on this construal of it. (For future reference I'll call the first order construal of platonistic gravitational theory P_0 and the second order construal P.)[5] Indeed, with a bit of minor dickering we can take P to be an extension of N, in which case it is a conservative extension.[6] For P then we do get a conservative-sub-theory result. (Note that though the conservativeness of mathematics was used in proving that P has a conservative sub-theory, it was by no means sufficient for getting this result. The appropriate nominalistic theory N had to be formulated independently, and the extended-representation theorem had to be proved independently.)

There are various reasons (one of which will be mentioned in section 4) why I was (and am) discontent to rely on the complete logic of the *part of* relation. What happens if we try to get a nominalistic version of gravitational theory within a purely first order logic? There is a rather natural first order sub-theory of N to focus on, which I call N_0 in the book. When I first began to work seriously on the book, I hoped to establish that, though the representation theorem (1) can't possibly hold for N_0 (or any other first order theory), still N_0 was a conservative sub-theory of P_0.[7] My idea was that, just as the Skolem-Löwenheim and compactness theorems show that N_0, being first order, has nonstandard models, these theorems also show that P_0, being first order, has nonstandard models and in particular that there are nonstandard models of the set theory that P_0 contains.[8] So instead of looking for a representation function ϕ onto the quadruples of real numbers, why

[5] P_0 really hasn't been specified definitely until the axiomatic set theory in it has been specified definitely, and different specifications of the set theory produce slightly different nominalistic consequences. (This is not so for P: no consistent strengthening of a certain 'core' second order set theory produces any new nominalistic consequences in the gravitational theory.) I am imagining that the axiomatic set theory in P_0 be specified in some fairly typical way.

[6] The minor dickering involves eliminating the empty class and identifying each singleton class with the object that falls under it. (Here, 'class' means 'object in the range of the second order variables'. One may do analogous dickering for sets (which are among the objects of the first order variables) if one likes, but there is no need to.) This modification makes classes formally like regions: indeed, one can think of them as the regions of a fictitious platonic space whose points include not only points of physical space but sets as well.

[7] At least for some reasonable choice of set theory and hence of P_0: see fn. 5.

[8] The set theory that P_0 contains is a set theory with urelements in which there is a set of all urelements and in which the non-mathematical predicates of P_0 are allowed to appear in the comprehension axioms (i.e., in the instances of the separation or replacement schemas).

not look instead for a representation function ϕ onto the quadruples of entities that serve as real numbers in some (standard or nonstandard) model of P_0? (This would be a representation onto the quadruples of members of some real closed field of a very special and not easily determinable sort.) Similarly for the representation functions ρ and ψ. If it could be established that for each model of N_0 there was at least one model of P_0 such that there were representation functions ϕ, ρ and ψ from the space-time of the first model to the real numbers of the second, this would suffice to establish a weaker variant of the extended-representation theorem that would be all that was needed to establish that N_0 had P_0 as a conservative extension. And this is what I (rather naively) hoped to establish when I first began to work seriously on *Science without Numbers*. Unfortunately, my hopes of proving this were dashed when John Burgess and Yiannis Moschovakis pointed out to me that a Gödelian construction can be carried out in N_0. This yields a nominalistic sentence undecidable in N_0 but provable in P_0; so N_0 can't possibly be a conservative sub-theory of P_0. (Indeed, the same argument extends to show that P_0 can't have *any* 'natural' nominalistic sub-theories over which it is a conservative extension.)[9]

This result, which I reported in the last chapter of my book, is worked out more fully by Shapiro. (It is one of two closely related Gödelian results that he uses in his paper. Earlier in the paper he used Gödel's theorem in his demonstration that mathematics is not proof-theoretically conservative in the context of second order logic; I have discussed that fact in section 1, but it is of no relevance in the present context. What is now under discussion is Shapiro's second use of Gödel's theorem, on pp. 529–30 of his paper.) Shapiro uses this result as the basis for his assertion that my account of the application of mathematics to the physical world is undermined if the physical theories are taken to be first order. I tried to argue against this assessment in the last chapter of my book, but Shapiro argues in effect that what I say there is unsatisfying since it gives up on representation theorems which are essential to an account of how mathematics is applied to physics.

I will save the philosophical discussion of Shapiro's objections until sections 5 and 6. But in thinking abut this objection, it occurred to me that there *is* an extended-representation theorem not far below the surface of what I say in the book (though of course not quite so nice a one as those whose impossibility I just noted); and it will facilitate

---

[9] It can have unnatural ones, even some that are recursively axiomatized: for as remarked, the Craig reaxiomatization of the set of nominalistic consequences of P_0 is certainly a conservative subtheory of P_0. This Craig reaxiomatization will of course have a Gödel sentence, but the Gödel sentence won't be provable in the set theory used in P_0.

the philosophical discussion in sections 5 and 6 to state it here. The modified extended-representation theorem says that, for any model of N_0, there is some real closed field F and some subset S of the power set of F^4 (the set of quadruples of members of F) with the following properties:

(i) $\langle F^4, S \rangle$ forms a Henkin model of an expanded version (which I will call Q) of the second order theory of quadruples of real numbers;[10]

(ii) (1_F) holds, where (1_F) is just like (1) except that

        (a) 'real numbers' is replaced throughout by 'members of the field F', and

        (b) 'laws of Newtonian gravitational theory in their functional form' is replaced by 'standard axioms of

---

[10] By a Henkin model of the *usual* second order theory of quadruples of reals, I mean an ordinary model of the following theory T. T is a first order theory in a three-sorted language; call the three sorts of variables the *real number variables*, the *quadruple variables*, and the *set variables*. For the number variables, assume the axioms for ordered fields. Use a 4-place functor $\langle\rangle$ from numbers to quadruples and four 1-place functors π_i from quadruples to numbers, defined everywhere, with the laws $\pi_i\langle x_1,\ldots,x_4\rangle = x_i$ and $\langle \pi_1(q),\ldots,\pi_4(q)\rangle = q$. For the sets, assume that they have only quadruples as members and that there are enough sets to obey the comprehension schema

$$(^*)\quad \text{There is a set of all } q \text{ such that } F(q)$$

where F is any formula of the three-sorted language whose only vocabulary is the language of ordered fields together with $\langle\rangle$, the π_is, and ϵ. (The formula may contain free variable parameters.) Finally, assume a least-upper-bound axiom; e.g., by saying that, for each set S that contains at least one quadruple of the form $\langle y, 0, 0, 0\rangle$, if there is a number x such that whenever $\langle y, 0, 0, 0\rangle$ is in S then $y \leq x$, then there is a smallest such x. (In general we can identify real numbers with quadruples of reals of the form $\langle y, 0, 0, 0\rangle$; this will allow us to say things about sets of reals as well as about sets of quadruples.)

That is (a first order transcription of) the *usual* second order theory of quadruples of reals. To get (a first order transcription of) the expanded version I have in mind, we add a fourth sort of variables, namely, variables for physical entities (including space-time points and regions), and we introduce the usual physical vocabulary for talking about such entities, including numerical functors. (We can regard such functors as mapping physical entities into quadruples, given our identification of reals with certain quadruples.) The laws governing real numbers, quadruples and sets are the same as before, except that the new physical entity variables and the new vocabulary governing them are allowed to appear in the instances of $(^*)$. We are now allowed, for instance, to assert the existence of a set containing precisely those quadruples of real numbers which are space-time co-ordinates of points at which the gravitational potential is zero. (The effect of such new existence assertions on the model is easily seen to be invariant under those transformations of a co-ordinate system, etc., which are mentioned in (a)–(c) of (1).) These new set-existence assertions will lead to new assertions of the existence of real numbers and quadruples, because of the least-upper-bound axiom.

This expansion of the usual second order theory of quadruples is important: indeed, clause (b) of the weak representation theorem would make no sense without it.

> Newtonian gravitational theory as stated in their functorial form in Q'.[11]

This representation theorem is weaker than the attempted representation theorem discussed two paragraphs back, for this one does not claim that the pair $\langle F^4, S \rangle$ is isomorphic to the pair consisting of the things that serve as the quadruples of reals and the things that serve as the sets of such quadruples in some model of the set theory used in P_0. As a result, there is no guarantee that theorems about the reals stated in the language of Q, and provable in set theory but not in Q, will be true of $\langle F^4, S \rangle$. Consequently, the weak representation theorem does not guarantee that P_0 is a conservative extension of N_0; and, of course, we know already that it isn't.

The philosophical issue that the last part of Shapiro's paper raises, then, is the issue of whether this weak representation theorem is all that we need to account for the application of mathematics to the world in the Newtonian theory of gravitation. I will discuss this in sections 5 and 6.

But before discussing it, I want to say more about the 'logic' of the *part of* relation and about whether it is reasonable to use this logic in replying to the Quine-Putnam argument. To the extent that it is, the discussion in sections 5 and 6 will be unnecessary.

4

Many scientific theories have two features. First, they allow for the existence of infinitely many entities (e.g., infinitely many physical objects, or infinitely many parts of a single physical object, or infinitely

---

[11] Proof sketch: I refer in the book to a representation theorem by Szczerba and Tarski, which says that, for any model of a certain simple theory of affine geometry which refers to points only but not to larger regions, there is a representation function onto a four-dimensional affine space over a real-closed field. N_0 uses as its basis a stronger version of the Szczerba-Tarski theory of affine geometry, one which refers to regions and in effect contains the claim that any bounded region of a line contains closest bounds, plus a suitable schema asserting the existence of regions. Clearly this is enough to give the stronger requirements on F stated in the theorem I've cited. More importantly, it can be shown that this (together with the analogous 'closest bound' or Dedekind continuity axioms for mass density and gravitational potential) is enough to ensure that the field F that is used to co-ordinate space-time is order-isomorphic to the field used to serve as a scale for mass density and to the field used to serve as a scale for gravitational potential. (The notion of region that is common to the three closest bound axioms can be used to 'transfer' points in the field of one quantity to corresponding points in the fields of the others.) From these remarks it is easy to see that the technical work done in the book is enough to establish (1_F).

many parts of a light ray or infinitely many points or regions of physical space). Second, they utilize the relation 'is a part of' (or some other relation from which this can be defined) and indeed they allow that infinitely many entities are in the field of this relation. It is because Newtonian gravitational theory possesses these features that the problems raised in Shapiro's paper arise.

To see why this is so, reflect that one of the principles usually taken to govern the relation 'is a part of' can be put platonistically as follows:

(C_p) For any non-empty set S of entities (e.g., space-time regions) there is another entity u (e.g., another space-time region) which is the 'mereological sum' of the entities in S.

(For u to be the *mereological sum* of the entities in S is for u to have each of the entities in S as (not necessarily proper) parts, and to have no parts that don't overlap any of the entities in S; where for two entities to overlap is for them to have at least one part in common.) Now, it is clear that (C_p) is not a nominalistic assertion, for it speaks of sets. One might attempt to reformulate it as a schema:

(C_s) If $\exists x F(x)$, then there is a u such that u is the mereological sum of those entities y such that $F(y)$.

(It should be clear from the parenthetical definition following (C_p) that the phrase following the 'such that' in (C_s) can be stated in terms of 'part of' and F and quantification theory, without reference to sets.) Use of the schema (C_s) will give us arbitrary finite mereological sums, and any other mereological sums that are definable (with parameters) in the nominalistic theory in question; but it does not give all the mereological sums (the platonistically definable ones) that you get if you replace (C_s) by (C_p) plus a rich body of assumptions about sets. Of course it doesn't follow from this alone that there are platonistic theories that include (C_p) plus a rich body of assumptions about sets that have nominalistic consequences that you don't get when you replace (C_p) plus the set theory by the more meagre (C_s); but it explains how this could happen.

And it does happen: that is why the nominalistic theory N_0 falls short of the platonistic theory P_0 in nominalistic consequences. For the nominalistic theory N_0 contains (C_s) as a part; and it can easily be shown (as a consequence of the proof of the extended-representation theorem (1)) that if you replace (C_s) by (C_p) plus axiomatic set theory, the resulting non-nominalistic theory T is logically equivalent to P_0. Given this fact, the Gödelian result shows that, in the context of the rest of N_0, the passage from (C_s) to (C_p) plus axiomatic set theory does yield new nominalistic consequences.

In evaluating this situation it is important to realize that by shifting from (C_s) to (C_p) plus axiomatic set theory, one does not achieve a complete theory of the part-of relation (even in conjunction with the other axioms on that relation that appear in N_0 or in T). (C_p) plus axiomatic set theory does not guarantee the existence of all regions that intuitively exist, any more than (C_s) does, because of the existence of nonstandard models that don't contain 'all' sets. These nonstandard models can be exploited to produce undecidable sentences in the theory T built on (C_p) plus axiomatic set theory, in the same way that the nonstandard models of (C_s) can be exploited to produce undecidable sentences of N_0. (For instance, the Gödelian construction used for N_0 can also be applied to T or, equivalently, to P_0.)

One could of course insist that one knows more about sets than is given in axiomatic set theory – that one has an intuitive grasp of the standard model for set theory – and that in evaluating what follows from a theory such as T which contains (C_p) one ought consider only models of that theory in which the sub-model assigned to the set theory is standard. But if this is legitimate (which, by the way, I doubt), then why not do the same for regions rather than for sets? That is, why not insist that one knows more about regions than is given in the axiomatic region theory that includes (C_s), for one has an intuitive grasp on the standard model of mereology; so that in evaluating what follows from N_0 one ought consider only those models of N_0 in which the sub-model assigned to the theory of regions is standard? Such a restriction to the standard model of mereology can't be any more philosophically dubious than a restriction to the standard model of set theory; and it can be imposed without characterizing regions via extrinsic entities (sets) in the manner of (C_p).[12]

The idea that one should restrict one's attention to the standard model of mereology in evaluating what follows from N_0 is simply another way

[12] One of the journal editors has objected (a) that the intuitive basis of the complete theory of the part-of relation is derivative from the idea of a set, and (b) that, if so, this undermines the claim in the text. I disagree on both counts. Indeed, as far as (a) goes, I suspect that the 'order of intuitions' is the other way around: set theory has the theory of the part-of relation as its main intuitive basis, and extends beyond this intuitive basis in various ways: adding an empty 'region' or set; distinguishing a point (smallest region) from a set containing that point; and adding sets of higher rank that don't correspond to regions at all. As for (b), I concede that the claim in this paragraph of the text would be undermined by the thesis that one can have an intuitive grasp of the standard model of set theory independently of mereological ideas, but that one can intuitively grasp the standard model of mereology only via an intuitive grasp of the standard model of set theory. But of course that thesis is very much stronger than (a) (which makes no mention of an intuitive grasp of standard models) and I see no reason why anyone should believe the stronger thesis.

of saying that one should shift from N_0 to the theory N which uses 'the complete logic of the part-of relation'. I have not claimed that doing this is philosophically defensible; only that it is *as* philosophically defensible as the restriction to the standard model of set theory in evaluating the consequences of platonistic theories (which is in effect a shift from first order platonistic theories to second order platonistic theories).

One source of doubt about the 'logic' of the part-of relation is that it has no complete proof procedure. A number of readers of *Science without Numbers* have been puzzled as to how the appeal to the relation of semantic consequence in a logic without a complete proof procedure can possibly be of use. This does seem puzzling at first, since surely it seems plausible that the means by which we can know that one thing follows from another are codifiable into a proof procedure, and that seems to imply that no appeal to anything stronger than the proof procedure can be of practical utility. (Perhaps such an argument is part of the reason why Shapiro thought I would be better off talking about derivability rather than semantic consequence in defining conservativeness.) But a closer look at the way in which the logic of the part-of relation is used will show that there is no need to attribute to the user any epistemological access to more than a recursively enumerable part of it.

To see this, let us ask what is required for a person to have good reason to believe that a nominalistic claim A is a semantic consequence of the nominalistic theory N that uses the logic of the part-of relation. If such a person were to proceed directly, i.e., without using mathematics, it would presumably be required that he or she be able to derive A from N using some proof procedure D for (a fragment of) the logic. But how about if the person appeals to a mathematical theory M and to semantic conservativeness? In that case, the person must have good reason to believe two things:

(i)   that A is a semantic consequence of N+M; and

(ii)   that M is semantically conservative (or at least, that it is semantically conservative *over N*, i.e., that N+M is a semantically conservative extension of N).

Now, in order to have good reason to believe (i), one must presumably be able to derive A from N+M using the proof procedure D. But note that, for this to happen, A need not be derivable from N alone using D, for the mathematics is not syntactically conservative. We see then that as long as one can have good reason to believe (ii), there is no difficulty in putting semantic conservativeness to practical use even for those very examples which give rise to a failure of syntactic conservativeness. (I believe that this and related facts are enough to undercut a great deal of the worry that one initially feels at the use of logics with no complete proof procedures.)[13]

There is still a philosophical question here, the question of whether one can reasonably believe (ii). A full discussion of this question would be far beyond the bounds of this essay, for it would involve issues about knowledge of metalogical relations that arise whether or not the logic in question has a complete proof procedure. But I would like to note that, if the goal is only to explain how we can reasonably believe that M is semantically conservative *over the particular theory N* (i.e., to explain how we can reasonably believe the weak parenthetical form of (ii), which is all that is needed in the present context), then the belief whose reasonability we must explain is a straightforwardly modal belief. For, since the part of N that describes the structure of space-time is categorical, it is easy to show that M is semantically conservative over N if and only if either N is semantically inconsistent or M is semantically consistent with N;[14] and this means in effect that

$$(2) \quad \Diamond AX_N \supset \Diamond (AX_N \ \& \ AX_M)$$

---

[13] It should be noted that the application of a typical proof procedure for (a fragment of) second order logic to N+M yields more than one gets by applying the proof procedure to N and to M separately and combining the results. For it yields not only a fragment of N and a fragment of M, but also the 'mixed' statement (C_p) of several pages back.

[14] The proof is simplified if we suppose that, in N, physical objects are identified with the space-time regions they occupy. Then let B_N be any model of N and B_M be any model of N+M. Let B_M^S be the restriction of B_M to the domain of space-time regions and the vocabulary of space-time structure; let B_N^S be analogous, except that, by our simplifying assumption, the restriction of domain is vacuous. By categoricity, there is an isomorphism h from B_M^S to B_N^S. Let B have the same domain as B_M, but let the extension of the physical predicates be determined by B_N, in the sense that $\langle x_1, \ldots, x_n \rangle$ is to be in the B-extension of a physical predicate if and only if $\langle h(x_1), \ldots, h(x_n) \rangle$ is in the B_N-extension of that predicate. Then B is a model of M; for, since M is second order, B_M and, hence, B contain every set of space-time regions; so there is no danger that by making the 'nominalistic base' of B look like B_N rather than B_M we will require the introduction of new sets into the model. (The claim that B is a model of M also depends on the simplifying assumption: without that, B, as so far defined, wouldn't contain sets of physical objects. But the model is easily expanded to one that will work, if the simplifying assumption is dropped.) Moreover, the non-mathematical part of B is isomorphic to B_N. We see then that if N+M is semantically consistent (i.e., if it has a model B_M), then any model B_N of N can be expanded to a model B of N+M whose non-mathematical part is just like the original model of N. So if B_N is a counterexample to the claim that a nominalistic sentence A follows from N, then B is a counterexample to the claim that A follows from N+M. This shows that, if M+N is consistent, then M is conservative over N.

This reasoning can all be carried out in M. In carrying it out, we are making use of modal knowledge of the form $\Box(AX_M \supset \ldots)$; so it is really modal knowledge of that form together with the modal knowledge that $\Diamond(AX_N \ \& \ AX_M)$, that I have shown to underlie the knowledge that M is conservative over N.

where '$\Diamond$' is an operator of logical possibility in a very strict sense, and where AX_N is the conjunction of the (finitely many) axioms of N and AX_M is the conjunction of the (finitely many) axioms of a standard version of second order set theory (such as Montague's). The real question, then, is: how reasonable is it to believe the modal claim (2)? Unfortunately, neither I nor anyone else that I know of has a great deal to say about the epistemology of modal claims; nevertheless, it seems pre-theoretically plausible to me that we can reasonably believe (2), and indeed that we can reasonably believe both its antecedent and its consequent.[15] I do not believe that there is any conflict between this claim and my nominalism; in particular, I would reject the idea that the only way to know that $\Diamond(AX_N \& AX_M)$ is to know that *actually* $AX_N \& AX_M$ (or to know that actually AX_M). Indeed, I would go further: I think it is no more problematic for a nominalist to claim that a belief that $\Diamond(AX_N \& AX_M)$ is reasonable than it is for a platonist to make this claim.[16] I discuss some of the issues involved in this paragraph further in 'Is mathematical knowledge just logical knowledge?'

[15] It seems to me *very* plausible that we can reasonably believe $\Diamond(AX_{N_0} \& AX_{M_0})$, where M_0 is first order set theory, though again I have no clear account of modal epistemology to support this plausibility judgement. Since second order set theory is a lot more powerful than first order set theory, it is *less* apparent to me that we can reasonably believe $\Diamond(AX_N \& AX_M)$; but it still seems fairly plausible that we can.

The plausibility of thinking that we can reasonably believe $\Diamond(AX_N \& AX_M)$ (or (2)) is sensitive to the way the second order set theory M is formulated. For instance, the continuum hypothesis is semantically decidable in second order set theory; if M' is M plus whichever one of $\{CH, \sim CH\}$ is a semantic consequence of M, then M' is logically equivalent to M (in second order logic), but I doubt that anyone could ever have very good reason to believe

$$\Diamond AX_N \supset \Diamond(AX_N \& AX_M')$$

This affects the epistemological situation discussed in the previous paragraph of the text. For if A is a nominalistic claim such that either it or its negation is derivable from N+M' by proof procedure D but such that neither is derivable from N+M by proof procedure D, then, since knowledge of (ii) is absent when M' is substituted for M, the epistemological discussion of the previous paragraph does not provide any means by which A or its negation could be known. I think this is as it should be.

[16] Of course, a platonist who believes the truth of AX_M can infer from this that $\Diamond AX_M$, and from that and some fairly uncontroversial assumptions about possibility it follows that $\Diamond(AX_N \& AX_M)$. But the belief that $\Diamond AX_M$ isn't made reasonable simply by following from some stronger belief (in this case AX_M) that entails it; that stronger belief must be reasonable.

The main doubts that arise about whether we can believe $\Diamond AX_M$ arise from the fact that there are variant second order set theories which can't be possible if M is possible: the challenge is to explain why it is more reasonable to believe $\Diamond AX_M$ than to believe $\Diamond AX_M^*$ (M* being such a variant). It seems to me that any reason that a platonist offers for believing that it is M rather than M* that is true can be taken over by a nominalist to argue with just as much force that it is M rather than M* that is possible.

I have been defending the use of the complete logic of the part-of relation against one (widespread) objection. Nevertheless I myself have substantial doubts about the philosophical legitimacy of regarding 'the logic of the part-of relation' (or any other form of second order logic) as genuinely a logic. For one thing, any such 'logic' violates a condition which I think ought to be required of a logic, namely, the condition of not making existential assertions. That is, the instances of (C_s) are existential claims; hence (on the view I incline toward) they shouldn't be regarded as logically true. But they are true in the standard model of mereology; so to adopt the 'logic' of the part-of relation by restricting attention to the standard model of mereology in evaluating consequence is to declare them logically true. Because of this doubt (and some others) I think that the extended-representation theorem (1) and the conservative-sub-theory result that follows from it probably do not ultimately yield a satisfactory reply to the Quine-Putnam argument, and that attention to the case of first order theories is required.

5

In these two final sections I want to argue that in order to reply to the Quine-Putnam argument that mathematics is indispensable to science and therefore should be regarded as true, we do not need a conservative-sub-theory result, nor do we need any representation theorem stronger than the theorem noted at the end of section 3.

Ultimately, of course, a discussion of the Quine-Putnam argument should be carried out not for Newtonian gravitational theory, but for physical theories that we regard as true. But such a discussion is not possible at present (since it must await advances in physics, and very likely in nominalization techniques as well). So, for the sake of argument, let us pretend that we have no evidence against Newtonian physics and that we have at least as much evidence for Newtonian physics as the evidence that was historically taken to support it. An evaluation of the Quine-Putnam argument requires that we evaluate whether, on this pretence, there are any advantages of platonistic formulations of Newtonian physics over nominalistic formulations, or vice versa.

Some of the considerations that go into answering this question are philosophical and of a sort that cannot be settled here; but where the nominalistic theory and the platonistic theory differ in their strictly physical (i.e., nominalistic) consequences, some of the considerations will involve these differences. The advantage of the situation where we have a non-trivial conservative-sub-theory result (e.g., the result that the second order theory P has a 'nice' second order nominalistic theory N over which it is a conservative extension) is that then no consideration of differences in the physical content of the two theories need be raised;

the philosophical advantages that one theory has over the other can be considered in a very pure form.

Suppose, however, that we want to compare two theories that differ in their nominalistic consequences. We will be particularly interested in the case where the nominalistic consequences of one theory are a proper subset of the nominalistic consequences of the other (as is the case for N_0 and P_0). The first question that arises is: is there any reason to believe in the truth of the excess consequences that one theory has but the other hasn't?

It seems to me that the answer to this question is that one should believe these excess consequences only to the extent that they arise from an explanatory superiority in the more powerful theory. And it seems to me that when the theories are of about equal overall simplicity (as are N_0 and P_0),[17] then an explanatory superiority in the more powerful theory can arise only if there is direct observational support for some of the excess physical consequences that it has and the weaker theory doesn't have. If this is right, then *even if we are platonists*, we have no reason to prefer P_0 to N_0 unless some of the consequences that P_0 has and N_0 lacks have direct observational support.

I can imagine this being disputed by a platonist, on the following grounds. P_0, we have seen, arises if we adjoin to N_0 both set theory and the assertion (C_p) that, for every set of points, there is a region containing exactly those points. But, from a platonist perspective, both set theory and the assertion (C_p) are true. Consequently, if we are platonists we ought to believe any excess physical consequences that we get from adjoining these truths to the rest of the theory, just as much as we believe the rest of the theory; consequently, if we are platonists, we should believe the physical consequences of P_0 if we believe the physical consequences of N_0.

Despite its superficial plausibility, this is a bad argument. (I'm afraid that the next to last paragraph of *Science with Numbers* comes close to endorsing it.) The reason that the argument is bad is that one of the axioms of N_0 (the Dedekind continuity axiom) says in effect that

(*) Any region of a line in space either is unbounded or has closest bounds.

---

[17] This remark is really over-generous to P_0. I think that, on any fair way of counting *simplicity of formulation* N_0 and P_0 come out extremely close, though P_0 may be slightly simpler in formulation than N_0. But even granting that there may be a slight advantage to P_0 over N_0 in simplicity of formulation, N_0 is much simpler than P_0 in ontology. A comparison of *overall* simplicity should take account both of simplicity of formulation and of simplicity of ontology; so I think that N_0 actually comes out much simpler than P_0.

To say this in the context of N_0 is in effect to say

(**) Any region that is nominalistically definable (with parameters) either is unbounded or has closest bounds.

Now if we were to formulate the axiom in the latter way and then add (C_p) plus set theory, we would get no new consequences for the structure of space: the regions that we add would be put to no use. But in adding (C_p) plus set theory to N_0 as actually formulated, the axiom (*), which had in effect meant (**), now becomes expanded into

(***) Any region that is platonistically definable (with parameters), either is unbounded or has closest bounds.

And to switch from (**) to (***) is not simply to assume the truth of (C_p) and mathematics; it is to make a substantive assumption about the structure of physical space which might not be true even if (**) and (C_p) and mathematics are all true. Consequently, even a platonist has no reason to assume the excess physical content that P_0 has over N_0 unless this excess content has some empirical support.

Now what I want to claim is that it is very hard to imagine any situation in which the excess content that P_0 has over N_0 would have empirical support; consequently, in any normal evidential situation, N_0 would be preferable to P_0 *independently of nominalistic scruples*; for it would be sufficiently powerful to explain all our observations in a satisfactory fashion without making the unsupported claims that P_0 makes.

In order to make a plausible case for this, it is essential that N_0 have all the *non-arcane* nominalistic consequences that P_0 has; for the non-arcane nominalistic consequences of P_0 are certainly well supported (under the pretence made earlier that we live in a basically Newtonian world). In *Science without Numbers* I conjectured without much explicit argument that the only nominalistic consequences on which P_0 and N_0 differ are the arcane consequences (such as the Gödel sentence of N_0). But I think that the extended-representation theorem noted at the end of section 3 gives support to the conjecture. For that theorem shows that whatever nominalistic consequences P_0 has that N_0 doesn't have must arise out of properties of the real number system which are provable in set theory but not provable in (i.e., not Henkin-consequences of) the second order theory of quadruples of real numbers. As far as I can tell, such properties of the real number system are never put to use in drawing consequences for which there is empirical support – indeed, as far as I know they are never put to use in drawing consequences which there is any clear way to put to experimental test. If this is right, then it is enough to establish the claims of the previous paragraph.[18]

6

The point of the preceding section can be put in a different way. We know that the platonistic theory P_0 has no natural conservative-sub-theory; but there is a weak variant of P_0, call it P_0^-, which does have a natural conservative sub-theory viz., N_0.[19] And P_0^- is a *better* theory than P_0, in that the extra empirical content that P_0 has over P_0^- has no empirical support and isn't even very obviously testable. P_0^- is like P_0 except that, instead of claiming that the structure of space-time is that of a four-dimensional Galilean space over the field of real numbers, it claims merely that the structure is that of a 4-dimensional Galilean space over some 'nice' real-closed field which might or might not be the field of real numbers. Here, a *nice real-closed field* is a field F for which there is a subset S of F^4 for which $\langle F^4, S \rangle$ is a Henkin model of Q; equivalently, it is an ordered field F in which each set of points that is definable (with parameters) in Q has a least upper bound. (P_0^- also says that the range of values of a continuous scalar quantity like gravitational potential is an interval of this same field F, rather than an interval of real numbers.) I grant that the restriction of the least-upper-bound axiom to sets definable in Q may appear a bit *ad hoc* when P_0^- is stated on its own independently of N_0. But it loses its slightly *ad hoc* character when it is realized that the restriction springs naturally from the fact that P_0^- is generated from an underlying nominalistic theory by means of a representation theorem.

[18] It is worth remarking that there is no law requiring a nominalist to make do with N_0: if it turns out that there is something true and important that P_0 contains but N_0 doesn't, then a nominalist is free to invoke a broader nominalistic theory N_1 that incorporates this extra claim. (N_1 might be a sub-theory of P_0; or it might contain important claims that P_0 doesn't contain but which can be empirically supported. It might even contain empirically supportable claims that conflict with part of the 'excess content' that P_0 has over N_0.) Indeed, it isn't clear that a nominalist who dislikes use of the second order theory N needs to accept a restriction to purely first order theories. One natural idea which I mention in my book is to use the quantifier 'there are only finitely many', to get a theory N^* which (though not purely first order) doesn't have some of the special problems that use of the logic of the part-of relation has. (Use of this new quantifier would allow us to get a representation theorem in which the real-closed field F was guaranteed to be Archimedean.) However, I think that the extended-representation theorem in section 3 is probably strong enough to show that there is no practical need for adding the finiteness quantifier or strengthening N_0 in any other (even purely first order) way.

[19] I hope it is clear that this does *not* imply that the theory of real numbers used in P_0 isn't conservative. If that is not clear, reread section 2.

At the end of section 3, I stated Shapiro's worry about the first order version of my nominalization strategy as a worry about whether the account of the application of mathematics to science which I give in my book requires a strong extended-representation theorem (i.e., a representation theorem of one of the sorts shown impossible for natural first order theories). According to the account offered in *Science without Numbers*, we apply mathematics by using it to associate with various nominalistic statements some platonistic statements that are called *abstract counterparts* of the nominalistic statements, and by then making inferences among the abstract counterparts which we transfer back to the nominalistic statements of which they are counterparts. (I argued there that the legitimacy of this procedure does not depend on the truth of the mathematics but only on its semantic conservativeness.) Does what I said in the previous section of this essay alleviate the worry that this account of application requires a strong representation theorem? Perhaps not entirely; for the account as just sketched turns on the use of the abstract counterpart relation. To say that a platonistic statement A is an abstract counterpart of a nominalistic statement B is to say that the material equivalence of A with B can be proved from the nominalistic physics together with standard mathematics. In most of the cases of interest, the proof that a given platonistic statement is an abstract counterpart of a given nominalistic statement relies on representation theorems; and this consequence of the representation theorems doesn't follow from the fact that the platonistic and nominalistic theories have the same or close to the same nominalistic consequences.

Actually it is not clear to me that the claim that a given platonistic statement is an abstract counterpart of a given nominalistic statement is of any great interest in its own right – that is, of any great interest except as a step in the proof that the nominalistic and platonistic theories have pretty much the same nominalistic consequences. But there is no need to discuss whether this is so, for in any case the weak representation theorem given at the end of section 3 is all we need to establish a large body of abstract counterpart claims. The only difference between the situation with N (where we have the strong representation theorem (1)) and the situation with N_0 is that, in the latter case, we cannot strictly assume that the abstract counterparts of nominalistic statements are statements that involve the real numbers. Rather, the abstract counterparts are statements about some undetermined nice real-closed field which may or may not be the real numbers. But this difference will not normally make any difference, for the nice real-closed fields are very much like the real numbers in all respects that are practically relevant: any typical calculation valid for the reals will be valid for all other nice real-closed fields as well. To put it more precisely, any calculation performable in the expanded second order theory Q of quadruples of real numbers holds for arbitrary nice real-closed fields (simply by the definition of niceness); so as long as we restrict ourselves to such calculations, no error can arise from assuming that the field in question is the field of real numbers. Consequently, the explanation that I offer in my book for why the real numbers are legitimately usable in facilitating inferences from nominalistic theories carries over. In addition, we are provided with a warning, which in the previous section I argued to be salutary: don't assume that the arcane properties of real numbers (those which aren't consequences of the theory Q) will be mirrored in the space-time of the physical world.

5

Platonism for cheap? Crispin Wright on Frege's context principle

Wright's new book (Wright 1983) is a sustained defence of arithmetical platonism. At the centre of Wright's defence is a version of Frege's famous dictum that 'only in the context of a sentence does a word have any reference.' Wright thinks that this dictum, properly understood, goes a long way toward establishing arithmetical platonism (and indeed, platonism about a large array of putative abstract objects). The book is impressive in the scope of considerations it involves and in the depth with which may important issues are pursued, and I suspect that many readers will find its defence of platonism persuasive. Despite the book's many virtues, it seems to me that the central argument is deeply flawed. It is that central argument that I will concentrate on in this critical notice.

1

What is the 'arithmetical platonism' that Wright is defending? It is the thesis that *numbers are objects*, where objects are opposed to Fregean concepts. (I'm not sure how much of Frege's rather peculiar doctrine of 'unsaturatedness' is supposed to be built into the notion of a 'Fregean concept'; I suspect that Wright would not regard any view according to which numbers are properties as platonistic in his sense.)

This thesis that numbers are objects can be usefully divided into two parts:

(1) Numbers, if there are any, are objects;
(2) There are numbers.

In defending platonism, Wright artfully intertwines considerations relevant to (1) with considerations relevant to (2), a practice that could make it seem to a careless reader that various considerations adduced in support of (1) support (2) as well. But I think that clarity is best achieved by disentangling them.

Actually I will be less concerned here with claim (1) than with another claim that Wright makes in arguing for it, namely

> (1′) Numerical phrases like 'the number 4' and 'the number of books in this room' function semantically as genuine singular terms and 'natural number' functions semantically as a predicate, indeed, a sortal predicate (i.e., a predicate the understanding of which determines the identity conditions of any object that falls under it).

I take (1′) to be compatible with the denial of (2), for I take it that a term's not denoting does not prevent it from functioning semantically as a genuine singular term, and that a predicate's having no extension does not keep it from being a sortal: 'Homer' functions semantically as a singular term whether or not there was anyone for whom it stands, and 'tachyon' and 'unicorn' are sortals even if there are no objects falling under them. The relation of (1′) to (1) is less clear. Frege, of course, thought that (1) follows from the first half of (1′), for he thought that no singular term could stand for a Fregean concept: even 'the concept of a horse' doesn't stand for a concept, but for an associated object. Wright gives a qualified defence of Frege's inference from (1′) to (1), but I would rather not get into those issues. Most of Wright's discussion of (1) is in fact concerned with (1′), and it is the issues involving (1′) (in addition to those involving (2)) that I will focus on.

Wright's argument for (1′) can be construed as having two steps. First, there is a syntactic argument for the claim

> (0) Numerical phrases like those mentioned in (1′) function syntactically as singular terms, and 'natural number' functions syntactically as a sortal predicate.

Second, Frege's Context Principle (which I will discuss in section 2) is then used to argue from (0) to (1′) (and ultimately to (2) as well).

Despite the apparent obviousness of claim (0), Wright's discussion of it is of considerable interest.[1] In the first place, he points out that it is less trivial than it might seem. On a trivial construal, the phrase 'John's sake' and 'the sake of the children' would also count as singular terms by syntactic criteria, and the word 'sake' would count as syntactically a sortal. But on Wright's view, this trivial construal is much too superficial, for it neglects the syntactic consideration that one cannot

---

[1] Or, the part of his discussion which in my reconstruction is relevant to claim (0). My numbered theses do not actually appear in Wright's book in so many words.

formulate intelligible identity statements involving sakes. The point of claim (0) is that 'the number of books in the room' satisfies *all* syntactic criteria for being singular terms, not just the most superficial ones. In the second place, Wright points out that there is another sort of numerical expression whose apparent syntactic behaviour is more complicated than that of descriptions like 'the number of books in the room': namely, numerals like 'three'. In some contexts, 'three' does not appear to be a singular term. For instance, in the sentence

(i)   There are at least three apples in the house

it appears to be part of the numerical quantifier expression 'there are at least three'; and it is not easy to reconcile this with its being a singular term. In other contexts, such as '$2+3=5$', it does appear to be a singular term. Wright seems to assume (pp. 10–11) that we need a single syntactic construal of all occurrences of numerals. The one he favours makes all occurrences into single terms: even 'three' in sentence (i) is a singular term. How can this be? The view is evidently that despite superficial syntactic evidence to the contrary, this sentence does *not* contain a numerical quantifier. Properly parsed, the numerical quantifier disappears and is replaced by the numerical functor 'the number of' plus the singular term 'three':

(i')   The number of apples in the house is greater than or equal to three.

The relevance of this to (0) is that someone might think that if some occurrences of numerical expressions are to be 'reparsed', it shouldn't be those which *prima facie* are parts of numerical quantifiers; it should be those which *prima facie* are singular terms. Wright discusses the possibility of reparsing the apparent numerical singular terms in terms of numerical quantifiers, and finds it unpromising. I am inclined to agree. My own view is that neither the singular terms nor the numerical quantifiers need reparsing.[2] The difference between this view and Wright's will be of some slight importance later. But the main lines of the issue of whether (0) leads to (1') and thence to (1) and (2) don't depend on whether the numerical phrases mentioned in (0) include those disputed occurrences of numerals which 'in surface structure' appear to be part of quantifier expressions.

---

[2] We must of course be able to explain the connection between (i) and (i'), but this can be done without attributing to 'three' the same logical form in both sentences. (For instance, one could hold that 'three' is a singular term in (i') but not in (i), but that (i') and (i) are logically equivalent. This is not the only possibility, however, and I will ultimately be defending another one.)

Now let's turn to (1′). One gets a sense of what (1′) comes to by considering a view which gets a great deal of attention in the book, and which Wright calls *ontological reductionism*. Consider those apparent singular terms of the form 'the direction of c', where 'c' stands for a line. Syntactically, these certainly appear to be singular terms. But according to the 'ontological reductionist' they do not function *semantically* as singular terms: for sentences that contain them are logically equivalent to sentences that make no reference to directions at all, but speak only of lines. For instance, the ontological reductionist holds that

(a) The direction of c_1 = the direction of c_2

is logically equivalent to

(a′) c_1 is parallel to c_2;

and that

(b) The direction of c_1 is closer to that of c_2 than to that of c_3

is logically equivalent to

(b′) c_1 is more nearly parallel to c_2 than to c_3.

The truth of the second member of each of the logically equivalent pairs does not require the existence of directions; so the truth of the first member of each of the logically equivalent pairs can't require the existence of directions either, and so phrases like 'the direction of c_1' can't be functioning semantically as singular terms. That is 'ontological reductionism' about directions.

Wright points out, quite correctly I believe, that there is something very peculiar about ontological reductionism as so stated. If one judges by syntactic criteria, 'the direction of c_1' and 'c_1' are both singular terms. If we assume that because 'c_1' is syntactically a singular term, then it must be semantically a singular term as well, then we can apparently conclude that (a′) and (b′) can be true without commitment to entities other than lines. If we also assume that sentences like (a) and (b) are logically equivalent to sentences like (a′) and (b′), we can then conclude that 'the direction of c_1' isn't semantically a singular term in (a) and (b) since these can be true without commitment to directions. But we could just as well have argued the other way: we could have assumed that since 'the direction of c_1' is syntactically a singular term it must be one semantically as well; and we could then have used the assumption of logical equivalence to argue that 'c_1' is not semantically a singular term in sentences like (a′) and (b′) since those sentences, by their equivalence to (a) and (b), require the existence of directions for their truth. This latter alternative, which might be called 'ontological inflationism', seems formally just as reasonable as ontological reductionism.

It should be noted that ontological inflationism and ontological reductionism are not theses about what exists: they are theses about which syntactic singular terms are semantic singular terms. (Claims about what must exist if certain sentences are to be true follow from each of the two theses.) Both of these theses are opposed to what might be called 'the face value thesis' according to which both 'line c' and 'the direction of line c' function semantically as singular terms. In my view, the face value thesis should be accepted. It seems to me that this *ought* to be Wright's view too for, as we will see in the next section, a large part of the content of the Context Principle on which he puts so much weight is that any expression which satisfies all the syntactic criteria for being a singular term ought to be regarded semantically as a singular term as well. Whether the face value thesis actually does represent Wright's view is not very easy to ascertain, and depends on the interpretation of some crucial passages (pp. 88–9 and 152–3) that we'll discuss later.

If one accepts the face value thesis, what is one to make of the putative 'logical equivalence' of (a) with (a')? The most obvious line to take (and the line that I would take) is that such claims of logical equivalence should be rejected: (a) can't be logically equivalent to (a') since (a) requires the existence of directions in order to be literally true, whereas (a') doesn't. This may appear paradoxical: (a) and (a') certainly appear to be intimately connected, and if the intimate connection isn't logical equivalence then what is it? I will return to this matter in section 3. For now, the only point I'm making is that there is at least one intelligible alternative to the view that some apparent singular terms need reparsing.

Although Wright's own presentation tends to obscure this, it should be clear that to oppose what Wright calls 'ontological reductionism' is not to deny that the 'equivalences' of (a) and (b) with (a') and (b') can be used to reduce ontological commitment. The point (Quine's point, at least as I have always understood him) is that the *sentences* (a) and (b) are ontologically committed to entities (namely, directions) that the *sentences* (a') and (b') are not ontologically committed to. Nevertheless, a *speaker who utters* (a) or (b) needn't be committed to these additional entities. For the ontological commitments of a speaker are the ontological commitments of the sentences he or she accepts in the strictest sense. A speaker who is aware of the fact that (a') and (b') are in some loose sense equivalent to (a) and (b) – e.g., that they serve all the useful purposes of (a) and (b) – will often utter (a) or (b) even though he or she doesn't accept it in the strictest sense but only accepts (a′) or (b′). If this is the way we view reduction of ontological commitment it seems gratuitous to accept the 'ontological reductionism' that Wright spends so much time arguing against.

'Ontological reductionism' for numbers seems to me even less plausible than for directions: for in the case of numbers, there is no generally agreed on procedure for how numerical singular terms should be eliminated in all occurrences – especially if we insist that the elimination does not introduce further elements (such as classes) of equally dubious status. Assuming, then, that ontological reductionism is to be abandoned, there seems to be little motivation once we have granted (0) for refusing to grant (1′) as well; and I am happy to concede for present purposes that (1) follows from (1′) plus other reasonable assumptions.

The real guts of arithmetical platonism, however, is (2). Let us see how Wright proposes to argue for it.

2

Wright's argument for (2) is based heavily on (a quite plausible exegesis of) Frege's Context Principle (i.e., the famous dictum quoted in my opening paragraph). According to Wright's exegesis, the principle involves two components. The first component is today pretty uncontroversial: it says that a satisfactory account of the meaning or reference of a subsentential expression must make clear its contribution to the meaning or truth conditions of sentences that contain it. The second component is more controversial: Wright labels it 'the thesis of the priority of syntactic over ontological categories' (p. 51). Wright says that this second component of the Context Principle explains Frege's platonism. Indeed, he says that it explains the fact that Frege was a *non-Gödelian* platonist. A Gödelian platonist is a platonist who explains our knowledge of mathematical entities in terms of some kind of quasi-perceptual relation to those entities. On Wright's view, a proper understanding of Frege's Context Principle (and in particular, of the Priority Thesis that serves as its second component) will not only show that there are lots of mathematical entities, it will show also that we don't need a Gödelian faculty of perception to explain our epistemological access to them.

What is this Priority Thesis that performs these functions? Wright introduces it as follows:

According to this reading, then, Frege is treating linguistic facts as decisive of whether or not a concept is genuinely sortal in the sense glossed earlier. Let us grant that the sort of considerations sketched go a good way towards establishing that the syntax of our numerical language can be very closely assimilated to that characteristic of talk of less controversial kinds of object [i.e., let us grant thesis (0).] . . . Frege's proposal, I suggest, is that the fact that our arithmetical language has these features is sufficient to set up natural number as a sortal concept, whose instances, if it has any, will thus be *objects*, furnishings of the world every bit as objective as mountains, rivers and trees. And, once again, that the concept does indeed have instances is settled by the truth of the appropriate arithmetical statements.

But how can such considerations be enough to settle the matter? For could it not, surely, simply be a *mistake* to give the syntax of our arithmetical language the kind of significance which Frege, on this interpretation, is proposing? *What if there really are no such genuine objects? And how, to reiterate the empiricist worry, can we possibly satisfy ourselves that there are such objects if there can be no empirical confrontation with them? Well, it is evident that Frege's position requires that such doubts be vacuous; there is to be no possibility of such a mistake* [my italics], no possibility that, the syntax of our arithmetical language and the truth of appropriate statements expressed in it notwithstanding, there are no such genuine objects. . .

Frege requires that there is no possibility that we might discard the preconceptions inbuilt into the syntax of our arithmetical language, and, the scales having dropped from our eyes, as it were, find that in reality there are no natural numbers, that in our old way of speaking we had not succeeded in referring to anything. Rather, it has to be the case that when it has been established, by the sort of syntactic criteria sketched, that a given class of terms are functioning as singular terms, and when it has been verified that certain appropriate sentences containing them are, by ordinary criteria, true, then it follows that those terms do genuinely refer. And, being singular terms, their reference will be to objects. There is to be no further, intelligible question whether such terms *really* have a reference, whether there really *are* such objects. (pp. 13–14)

There is much here that requires comment. First of all, note that the first quoted paragraph, minus the last sentence, says in effect that the Priority Thesis licences the derivation of (1′) (indeed, of (1)) from (0). Evidently then the Priority Thesis says at least this much: that any expression which by syntactic criteria counts as a singular term also functions semantically as a singular term. Let's call that the Weak Priority Thesis. It rules out 'ontological reductionism' of the sort discussed in the previous section.

The second point is that the Weak Priority Thesis is strong enough to have *conditional* implications about existence. Precisely what one takes these implications to be depends on one's views about the logic of singular terms that needn't denote. One common view – evidently Wright's view,[3] and for convenience the view that I will adopt here –

[3] See fn. 4 to Wright's section II p. 171.

is that existential generalization is valid for atomic statements. Equivalently, the view is that

(*) If F(t) then t exists

holds for any singular term t and any F for which 'F(t)' is atomic; here 't exists' abbreviates '$\exists x(x = t)$.' Consequently, we can infer the existence of the number 3 from the premise

(ii) 2 + 1 = 3,

or alternatively from the premise

(iii) The number of books in this room is greater than or equal to three.

(Similar inferences to the existence of the number 3 are valid on other reasonable views of the logic of terms that needn't denote; nothing of metaphysical substance hangs on Wright's and my choice of this one.)

However, it does *not* follow from the Weak Priority Thesis (together with syntactic claim (0)) that doubts about the existence of numbers are 'vacuous'. What does follow is that if we reject the existence of numbers we have to be fairly radical about it; we have to refrain from strictly accepting such claims as (ii) and (iii). But provided we are willing to be this radical, the Weak Priority Thesis is quite compatible with the denial of the platonistic thesis (2).

Furthermore, the 'radicalism' that the Weak Priority Thesis forces on the nominalist isn't really all that extreme. For instance, though the Weak Priority Thesis (together with the syntactic claim (0)) prevents the nominalist from strictly believing (iii), it does not prevent him or her from believing

(iii′) There are at least three books in the room,

where 'three' appears on the surface to be not a singular term but part of a numerical quantifier: nothing keeps the nominalist from accepting this unless we can argue on syntactic grounds that the surface structure of the sentence is misleading and that it must be reparsed in the syntactic form (iii). (Even were such a syntactic argument provided – which I very much doubt can be done – the nominalist could still believe

(iii″) There are x, y and z, all distinct, that are books in this room.)

Consequently, a nominalist may, consistently with the Weak Priority Thesis (and hence the denial of 'ontological reductionism'), regard (iii) as a strictly unacceptable way of communicating something that is perfectly acceptable, viz., (iii′) (or (iii″)). Of course, some story would still need to be told to explain why we bother asserting strictly unacceptable things like (iii) when we could have asserted acceptable things like (iii') instead; also, to explain why we assert strictly unacceptable things like (ii). Explaining these things would be part of making nominalism a plausible doctrine. Still, the fact remains that the Weak Priority Thesis (even together with syntactic assumption (0)) does not by any means preclude the nominalist position.

There is some question in my mind as to whether Wright intends the content of the Priority Thesis (i.e., the second strand of the Context Principle) to be exhausted by the Weak Priority Thesis. If this is intended, then I agree with Wright that Frege's Context Principle should be accepted, but do not see how it can have the implications that Wright thinks it has. For instance, I cannot see (to paraphrase part of the third paragraph of the passage quoted) how it can be 'a preconception inbuilt into the syntax of our arithmetical language' that '4' is not only a singular term but one which in fact denotes. Is it a syntactic presupposition of our historical language that 'Homer' denotes, or of our religious language that 'God' denotes? Are doubts about the existence of Homer and of God 'vacuous' for that reason?

It seems best to construe Wright as claiming more content to the Priority Thesis than merely the Weak Priority Thesis – but just what more is hard to state precisely. The best clue is offered in the next to last sentence quoted:

Rather, it has to be the case that when it has been established, by the sort of syntactic criteria sketched, that a given class of terms are functioning as singular terms, and when it has been verified that certain appropriate sentences containing them are, *by ordinary criteria*, true, then it follows that those terms do genuinely refer [my italics]. (p. 14)

The kicker here is the phrase 'by ordinary criteria': if it were omitted, we would have only the Weak Priority Thesis. I propose that the Strong Priority Thesis be given by the sentence quoted, with the italicized phrase taken very seriously. In effect, then, it is the Weak Priority Thesis plus the claim

> (S) What is true according to ordinary criteria really is true, and any doubts that this is so are vacuous.

Not that the content of (S) is clear in general. (Did the 'ordinary criteria' for truth in ancient Greece make 'Zeus is throwing thunderbolts' true whenever there was lightning?) Nevertheless, it is clear how Wright intends it to be applied in the case of numerical statements like (iii): the 'ordinary criteria' for whether (iii) is true are stated in (iii') (or (iii'')). Similarly, in the case of directions the 'ordinary criteria' for the truth of the claims (a) and (b) are stated in claims (a′) and (b′). On this construal of 'ordinary criteria', claim (S) of the Strong Priority Thesis strikes me as highly questionable (if, like Wright, we rule out 'ontological reductionist' reconstruals of apparent singular terms). The Zeus comparison seems apt: just because people ordinarily take (iii) as equivalent to (iii′) or (a) as equivalent to (a′) or 'Zeus is throwing thunderbolts' as equivalent to 'there is lightning', why assume that they must be correct? Certainly Wright offers very little in argument for claim (S) – indeed, the only argument for it that I have been able to find is given in one paragraph near the end of the book, a paragraph I will quote and discuss towards the end of section 4. Since the Weak Priority Thesis does not support platonism without claim (S), it seems to me that the impression which Wright's book gives of offering a positive argument for platonism is largely illusory. (The central argument of Wright's book can also be construed not as a positive argument for platonism but as a defence of claim (S) and platonism against an epistemological objection. Whether the argument is more successful in this guise is also something I will discuss in section 4.)

3

The doubts I have just raised about the Strong Priority Thesis may seem academic: surely, it may be said, claim (iii) about numbers and books is just *equivalent* to claim (iii′) about books alone, and claim (a) about directions is just *equivalent* to claim (a′) about lines; consequently, the ordinary criteria for claims (iii) and (a) are simply beyond doubt, in accordance with the Strong Priority Thesis. My response (already hinted at in my earlier remarks on reducing ontological commitment) is that there is indeed an important sense in which (iii) is equivalent to (iii′) and in which (a) is equivalent to (a′); but I do not believe that the sense in which they are equivalent is nearly enough to support the Strong Priority Thesis.

Before elaborating on this, it will be helpful to obtain a clearer sense of Wright's overall position on these 'equivalences', by formulating in rather general terms the sort of situation where Wright feels that the introduction of abstract entities is unproblematic. Suppose we have a theory N which contains a 1-place predicate L (e.g., 'line') and a 2-place predicate $\cong$ (e.g., 'parallel to'), and an axiom (E) guaranteeing that $\cong$ is an equivalence relation on the things that satisfy L. Suppose that N also contains further predicates $F_1, \ldots, F_n$ and further axioms $(C_1), \ldots, (C_n)$ guaranteeing that $\cong$ is a congruence with respect to (one or more designated term positions for) these predicates; e.g., if F_1

is a 3-place predicate (and all three term positions for it are designated as congruence positions), then the axiom (C₁) will say that

$$\forall x \forall x' \forall y \forall y' \forall z \forall z' \, (x \cong x' \, \& \, y \cong y' \, \& \, z \cong z' \, \& \, F_1(x, y, z) \supset F_1(x', y', z')).$$

(An example of such a 3-place F_k is the 'more nearly parallel to' predicate in (b').) N may also contain various other axioms $B_1, \ldots, B_m$ (possibly involving other non-logical vocabulary). Given such a theory N, it is natural to introduce a new theory A_N by 'abstracting on $\cong$'. A_N is to contain all the vocabulary of N, plus a new 1-place predicate D (e.g., 'direction'), plus a new 1-place functor τ (associating, e.g., directions with lines), plus new predicates $\phi_1, \ldots, \phi_n$ where each ϕ_k takes as many arguments as the corresponding F_k. The axioms of A_N will include

$$\forall x \, [L(x) \equiv \exists y \, (y = \tau(x))]$$
$$\forall y \, [D(y) \equiv \exists x \, (y = \tau(x))]$$

The axioms will not include (E), but instead

$$(D^=) \quad \forall x \forall y \, [L(x) \, \& \, L(y) \supset (\tau(x) = \tau(y) \equiv x \cong y)]$$

from which (E) follows. Nor will they include $(C_1), \ldots, (C_n)$, but instead things like

$$(C_1') \quad \forall x \forall y \forall z \, [L(x) \, \& \, L(y) \, \& \, L(z) \supset (\phi_1(\tau(x), \tau(y), \tau(z)) \equiv$$
$$F_1(x, \, y, \, z))]$$

(the exact formulation of each depending both on how many places F_k takes and which ones are designated as congruence positions); from these $(C_1), \ldots, (C_n)$ follow. Finally, for each additional axiom $B_1, \ldots, B_m$ of N, A_N will contain the same axiom except modified by restricting all quantifiers and free variables to things that do not satisfy the new predicate D. Let us say that A_N *arises from N by simple abstraction.*

The 'simple abstraction' model fits how our talk of directions is related to our talk of parallelism. As Wright emphasizes, it doesn't quite fit how our talk of numbers or sets is related to our non-numerical and non-set-theoretic talk; but Wright's idea is to model the introduction of numbers and sets on the introduction of directions, by starting from a theory N that contains not a 2-place *predicate* that represents an equivalence relation, but instead a 2-place *operator on formulas* which represents an equivalence operation (i.e., is reflexive, symmetric and transitive). For instance, sets are to be introduced from the operator 'exactly the same things are . . .s as are . . .s', via a predicate functor $\{x| \ldots\}$ (analogous to the τ of the simple abstraction case) which associates sets with (certain) formulas, and using the law

$$(\zeta^=) \quad \{x|A(x)\} = \{x|B(x)\} \equiv \forall x \, (A(x) \equiv B(x))$$

(Cf. 112).[4] It is assumed that the singular terms involving the new predicate functor $\{y|\ldots\}$ can be existentially generalized upon.[5] (This means that some restriction must be placed on the sort of formulas A for which $\{x|A(x)\}$ counts as grammatical, if paradoxes are to be avoided.) Similarly numbers are to be introduced from the binary quantifier 'there are exactly as many . . .s as . . .s', via a predicate functor η_x which associates numbers with formulas, using the law

$$(\eta^=) \quad \eta_x(A(x)) = \eta_x(B(x)) \equiv \text{ there are exactly as many } x \text{ such that}$$
$$A(x) \text{ as } x \text{ such that } B(x).$$

Again, it is assumed that the singular terms involving the new predicate functors can be existentially generalized upon. (See preceding footnote.) Wright thinks that in this case we do not need restrictions on the sort of formula A that can appear in the existentially generalizable singular terms $\eta_x(A(x))$: in particular, no paradox threatens even if they are allowed to contain arbitrary numerical vocabulary and to quantify over numbers without restriction. (This claim is very important to Wright's overall position, for reasons he explains well.) As far as I know, Wright is correct in thinking that no paradox looms in this case. (The fact that an apparently analogous form of abstraction *does* lead to paradox in the case of set theory should perhaps be disquieting to someone who wants to regard the principle $(\eta^=)$ as a logical truth, but I won't press this.)

Sliding over the details, it is evident that Wright's position about the introduction of numbers and sets is this: let N be any theory that contains a sufficient logic but that does not overtly postulate abstract objects. (Besides ordinary first order logic, the theory must include (if numbers are to be introduced) either a primitive binary quantifier 'there are exactly as many . . .s as . . .s' or else devices sufficient to define it. Wright's own choice is to define it via higher order existential quantification; this requires the assumption that higher order existential quantifiers do not assert the existence of abstract objects, which despite Wright's defence (pp. 132–4) seems to me either quite dubious or to

---

[4] It is important to bear in mind that on Wright's view sets as well as directions and numbers are to be introduced by abstraction. Otherwise, one may mistakenly suppose that when Wright talks about introducing directions or numbers by abstraction, he is assuming that sets are antecedently available and that the goal of the abstraction is simply to find some of these antecedently available sets to serve as directions or numbers. To make this supposition would be to totally misconceive Wright's project, as Wright says very explicitly (p. 150).

[5] This assumption follows from the policy on atomic statements discussed in section 2. (For from $(\zeta^=)$ we get $\{x|A(x)\} = \{x|A(x)\} \equiv \forall x \, (A(x) \equiv A(x))$; therefore $\{x|A(x)\} = \{x|A(x)\}$, and so by the policy on atomic statements, $\exists y \, (y = \{x|A(x)\})$.)

rest on an *ad hoc* distinction between entities in general and 'objects'. But this criticism doesn't cut against anything very central to Wright's programme, for he could either take the binary cardinality quantifier as primitive or follow Dale Gottlieb in defining it using substitutional quantification (with subscripts of numerical quantifiers as substituends).)[6] Given such an N, a theory A_N is to be introduced by abstraction on some equivalence operator such as 'exactly the same things are' or 'there are exactly as many' that appears in N; and A_N will include some law such as $(\zeta^=)$ or $(\eta^=)$ which introduces identity conditions for the new entities (and implicitly introduces existence conditions for them as well, by existential generalization). (Indeed, Wright argues that the theory A_N so introduced in the number-theoretic case will be sufficient to entail the usual laws of natural numbers (Peano's axioms).)

I think it is evident that the method of introducing abstract objects that is illustrated in the three examples of directions, sets and numbers is of considerable importance to an understanding of the function of abstract objects in our theorizing. It is a feature of this method that each type of abstract object is introduced as inseparable from a standard theory about that type of object (the various theories A_N). In the case of directions, the associated theory obtained by abstracting on 'is parallel to' can be called *direction theory*; in the case of numbers, we'd better call the associated theory obtained by abstraction *numerical theory*, since 'number theory' has already been taken. It seems clear that any use to which directions, numbers etc. are standardly put is going to involve much of the rest of direction theory, numerical theory etc. as well: certainly the bare existence of directions or numbers would count for little. So the issue of whether directions or numbers exist, and if so how we know it, is inseparable from the question of whether direction theory or numerical theory is true, and if so how we know that.

Now let us return to the status of the 'equivalences' between, on the one hand, platonistic claims like (a) and (iii) and, on the other hand, non-platonistic claims like (a') and (iii'). It should now be evident that there is a quite uncontroversial sense in which (a) *is* equivalent to (a') and (iii) to (iii'): (a) is equivalent to (a') *within direction theory*, in the sense that direction theory entails the material equivalence of (a) with (a'); similarly, (iii) is equivalent to (iii') *within numerical theory*. Since direction theory and numerical theory are important theories, intimately bound up with applications of directions and numbers, this is quite an important sort of equivalence. We have no business regarding it as a species of logical equivalence, however, unless we can find some argument for regarding direction theory and numerical theory as logically

---

[6] Dale Gottlieb, *Ontological Economy: substitutional quantification and mathematics*, p. 92.

true. *Prime facie*, the equivalences are *not* logical equivalences, since (a) and (a′) appear to differ in their ontological commitments, as do (iii) and (iii′); consequently, it appears *prima facie* that direction theory and numerical theory are not logically true. (To the extent that this *prima facie* argument is persuasive, it establishes not only that direction theory isn't logically true, but also that it isn't a logical consequence of *parallelism theory*, i.e., of the theory N of parallelism from which direction theory was obtained by abstraction.)

It might be objected that the importance of the equivalences between (a) and (iii) on the one hand and (a′) and (iii′) on the other requires, if not the assumption that direction theory and numerical theory are *logically* true, at least the assumption that they are *true*. But I disagree: one can explain why equivalence within direction theory and equivalence within numerical theory are important even while holding that direction theory and numerical theory are false.

Before explaining this, I want to separate clearly two problems that a defender of the view that direction theory and numerical theory are false faces. The first problem is the problem of defending the denial of numbers or directions against indispensability arguments of the sort that have been championed by Quine and Putnam. The problem, in other words, is to show that we can express anything we need to express, and explain anything we need to explain, without ever asserting the existence of numbers or directions. It is my view that in order for a denial of the existence of numbers or directions to be reasonable, this problem must be solved. Whether or not it can be solved is not an issue that I will address here; it is not a matter relevant to Wright's book, for Wright is proposing an argument for platonism that is clearly supposed to be independent of the Quine-Putnam indispensability arguments.

The second problem is the problem of explaining the importance of direction theory and numerical theory, and of the relations of *equivalence within direction theory* and *equivalence within numerical theory*, on the *assumption* that there are no directions or numbers and hence that direction theory and numerical theory are false. The following remarks are addressed to this second problem only, not to the first.

The idea is simple: it is that the transition from a theory N to the associated theory A_N is a transition from a theory to a conservative extension of that theory.[7] For instance, suppose we are interested in

---

[7] More accurately, A_N is a conservative extension of the theory N* that results from restricting all free variables and quantifiers of N by the clause 'is not a direction' (or 'is not a number', etc.). (N* is an agnostic version of N: N may presuppose in its unspecified laws that there are no directions (or numbers, etc.), and N* removes any such presupposition.) By the same token, 'parallelism theory' should really be taken as a name of N*, not of N. I have ignored the need of shifting from N to N* in the text only for simplifying the exposition.

claims in the language of parallelism, formulated without reference to directions (and with quantifiers explicitly restricted to non-directions – see preceding footnote); and suppose that we believe the laws of parallelism theory but disbelieve in abstract entities like directions. Even so, there is no harm in *feigning* to accept direction theory in making inferences among such claims: for any inference among such claims that is licensed by direction theory is licensed by parallelism theory as well.[8] The additional content that direction theory has over parallelism theory (e.g., the existence of directions) is *harmless* – at any rate, it is harmless *for the purposes of drawing consequences that say nothing about directions*. And since (a) is equivalent to (a′) within direction theory, we see that it is harmless to pass back and forth between (a) and (a′) as long as our only interest is in claims that say nothing about directions. If our interest is in ontology, of course, we are also interested in claims that do say something about directions – we're interested in claims that assert or deny that directions exist – so for *these* purposes the fact that direction theory is a conservative extension of parallelism theory is not enough to justify regarding (a) and (a′) as equivalent.

The same holds for entities such as sets and numbers that are introduced by the more complicated kind of abstraction (from operators rather than predicates), as long as we are careful to formulate the more complicated kind of abstraction in a way that doesn't give rise to paradoxes. For instance, if our interest is only with inferences among claims that don't say anything about numbers (but which may employ, say, numerical quantifiers), then we can employ numerical theory without harm, for we will get no conclusions with numerical theory that wouldn't be valid without it. Indeed, numerical theory is not only *harmless* for the purpose of drawing nominalistic consequences, it is also rather *useful* for this purpose: it is often easier to find (and see the correctness of) chains of inference from a body of nominalistic claims plus numerical theory than it is to find (and see the correctness of) corresponding chains of reasoning from a body of nominalistic claims alone. (To a much lesser extent, direction theory can also be useful as well as harmless.) Since (iii) is equivalent to (iii′) within numerical theory, we see that it is harmless (and indeed rather useful) to pass back and forth between (iii) and (iii′) when our interest is in conclusions that say nothing about numbers. There are other purposes for which this justification for feigning acceptance of numerical theory does not apply, and we must decide whether or not to genuinely accept the theory. For instance, there may be observations that we want to formulate that we don't see how to formulate without reference to numbers, or there may

---

[8] That is, if $p_1, \ldots, p_n$, q are such claims, and direction theory plus $p_1, \ldots, p_n$ logically entail q, then parallelism theory plus $p_1, \ldots, p_n$ logically entail q.

be explanations that we want to state that we can't see how to state without reference to numbers. No conservative extension results can justify our feigning to accept numerical theory for these purposes; if such circumstances do arise, then we will have to genuinely accept numerical theory if we are not to reduce our ability to formulate our observations or our explanations. (My own view is that such circumstances do not arise, but I will not argue for this here since, as noted, Wright (unlike Quine and Putnam) in no way bases his case for platonism on the supposition that they do arise.)

It is clear that we cannot use conservative extension results to justify treating (iii) as equivalent to (iii′) in contexts where our ultimate purposes are ontological. If our purposes are ontological we must decide whether to genuinely accept numerical theory (and decide also whether if we do accept it to regard it as logically true or as true for non-logical reasons). But whatever our decision on this, the fact that (iii) can in many contexts be regarded as equivalent to (iii′) is a fact that is easily explained even on the assumption that numerical theory is not only not logically true but actually false.

Wright, of course, thinks that (iii) and (iii′) are equivalent in a much stronger sense than this. In the next section I will inquire whether his position on this question can reasonably be maintained, especially given his denial of ontological reductionism.

4

A central tenet of Wright's approach to ontology is that one can and should be a platonist without endorsing Gödelian views about the 'perception' of abstract objects. If we understand the Context Principle correctly, Wright thinks, we see that there is no need for such a faculty of 'perception'.[9] For we can have knowledge of abstract objects if we can have knowledge of states of affairs in which those abstract objects figure. And (he maintains) knowledge of those states of affairs can be had without any ability to 'perceive' entities outside the physical realm.

This last claim is the crucial one, and Wright's main defence of it comes in one of the most important parts of the book (pp. 86–90). Wright begins as follows:

Imagine people who (as it was suggested earlier is possible) have mastered the use of both direction- and line-sentences by direct training in their use, without explicit mention of the relevant equivalences [e.g., of (a) with (a′)]; and suppose

---

[9] Nor, he apparently thinks, is there any need to maintain with Quine and Putnam that the existence of mathematical entities is supported in the same way as the existence of theoretical entities in science is supported: i.e., by an argument that they are indispensable in physical theory.

that these people come to see it as a problem to explain how knowledge of the truth or falsity of direction-sentences is possible, in view of plausible causal constraints on knowing and the acausality of directions *qua* abstract objects. Then, if they reflect and hit upon the equivalences – no very difficult matter – are they not in a position to give a perfect answer? Namely: knowledge about directions, admittedly abstract objects, is possible because statements about directions are made true or false by concrete states of affairs; it is in the properties and relations instantiated among lines that the truth or falsity of statements concerning direction consists. The answer, however, does *not* work by showing how knowledge "about" that kind of abstract object is really no such thing, and how these peoples' talk of directions is a *façon de parler*. The explanation cannot work that way for the reasons elaborated in the previous section, since their understanding of talk of directions was bestowed *independently* of knowledge of the relevant equivalences. (p. 87)

The last two sentences of this passage disavow ontological reductionism, i.e., they maintain that the semantics of direction statements are to be taken at face value. The rest of the passage maintains that, nevertheless, direction statements like (a) are made true or false by ordinary concrete states of affairs. The problem is to figure out how these two things can be reconciled.

In our earlier discussion we noted two possible alternatives to ontological reductionism. One of them, 'ontological inflationism', claimed that the surface grammar of claims like 'line c_1 is parallel to line c_2' is misleading: the real logical form of such a sentence is better expressed by 'the direction of line c_1 is identical to the direction of line c_2', which shows clearly the ontological commitment to directions. Now, ontological inflationism is not very plausible, but beyond that, it does not seem to accord with the early part of the above quotation. For what it says is that even a statement like (a′) is not made true by a concrete state of affairs involving lines only; rather, the state of affairs that makes it true, like the state of affairs that makes (a) true, involves abstract entities. Why then don't we need a Gödelian faculty of perceiving abstract objects to ascertain that there are abstract objects with the properties stated by (a) and (despite appearances) by (a′)?

The other possible alternative to ontological reductionism that I have mentioned is one which denies that (a) and (a′) are logically equivalent. But it is even clearer that this alternative does not support the early part of the above quotation. If (a) and (a′) aren't logically equivalent, then the concrete state of affairs that makes (a′) true does not suffice to make (a) true; so why don't we need a Gödelian faculty (or some other controversial means of epistemological access) to verify the excess content that (a) has over (a′)?

The difficulty, then, is to figure out how it is to be maintained that (a) can be verified by a concrete state of affairs without adhering to ontological reductionism. Wright recognizes that there is an apparent problem here, and responds to it as follows:

The best response would be, I think, to persevere with the strategy already outlined [in the passage quoted above], denying that it needs to rely upon a 'face-value' construal of [(a′)]. But the presentation of the strategy must be altered. It will no longer do to present the inter-translatability of direction- and line-sentences, for instance, as demonstrating how statements concerning abstract objects can be made true or false by concrete states of affairs; for the fact that line-sentences contain only concrete vocabulary can no longer straightforwardly be taken as an indication of the through-and-through concrete character of the relevant states of affairs. Rather, the emphasis must now be upon the *epistemologically unproblematic* character of statements about lines, upon the fact that straightforward practical operations suffice in general to verify or falsify them. The change is small but vital: the point of reduction of direction-sentences to line-sentences is not to demonstrate the causally unproblematic character of our knowledge of the former, but to remind us of – or bring to our attention – the existential presuppositions of the abstract tacitly present in familiar empirical statements whose knowability one would naturally have supposed to be uncompromised by any reasonable causal constraints upon knowing. (p. 89)

The most obvious way to interpret this is that Wright is advocating ontological inflationism, and saying that on the ontological inflationist position *all* states of affairs involve abstract objects, so that (a) is made true by states of affairs that are *as concrete as one can get*. How satisfactory would such a response be? It is certainly true that if one were convinced of the ontological inflationist position, one would have an extremely *strong motivation* to believe that we can have knowledge of states of affairs involving abstract objects, since the position that we can't have such knowledge would for the ontological inflationist be tantamount to general skepticism. It is not at all clear, however, that the fact that we would have strong motivation to believe the position that we can have knowledge of states of affairs involving abstract objects would do anything to show that the difficulties with the position could be overcome. But there is no need to pursue this, for in any case it is hard to seriously believe that Wright means to endorse ontological inflationism. Not only is it a wholly implausible semantic thesis, it is also a thesis that is in direct conflict with the Context Principle as Wright interprets it, according to which expressions that by syntactic criteria appear to be singular terms function semantically as singular terms.

The next interpretation I will consider is not entirely natural as an interpretation of the passage just quoted, but I think that it fits much better the overall theme of the book. According to this interpretation,

(a) is logically equivalent to (a′);[10] so that if we regard logic as epistemologically unproblematic, then (a) can be no more epistemologically problematic than (a′). This interpretation requires that (a′), though apparently about lines only, implies the existence of directions; but it maintains this compatibility with the face value thesis, by also maintaining that *the existence of directions is a logical consequence of the existence of lines.*[11] Of course, this claim should not be construed as saying that assertions of the form '$\exists x(x$ is a direction and $\psi(x))$' are to be understood as merely having the *appearance* of existential quantifications over directions, but as *really* meaning just '$\exists y(y$ is a line and $F(y).)$' *That* view would simply be an extended version of ontological reductionism: one that said that not only apparent singular terms for directions, but apparent existential quantifications over directions as well, must be viewed as semantically misleading. Wright is very clearly opposed to such an extended version of ontological reductionism; so if we are to interpret Wright as holding an alternative to ontological inflationism (as I think we should try to), then apparently the view must be this:

> (L) Directions are entirely acceptable entities distinct from, but with every bit as much reality as, physical lines; nevertheless, their existence (and their properties) follow logically from the existence (and properties) of lines.

The 'follow logically' in (L) apparently must be taken very literally; it is of course uncontroversial that the existence of directions follows from the existence of lines *in direction theory*, but that is not enough; nor would it be enough to add that direction theory is true. For it is essential to Wright's enterprise that he dismiss the epistemological worries that led Gödel to postulate a faculty that allows us to 'perceive' the abstract. That means that it is required that he hold direction theory to be not only true but in some sense *trivially* true; doubts about it are to be 'vacuous', which I assume means 'not ultimately intelligible'. And that seems to mean that the existence of directions follows from the existence of lines in a strictly logical sense.

Nominally, then, (L) provides an interpretation of Wright's views about the existence of directions. And it seems to be a good interpretation

---

[10] Or at least, (a) is a logical consequence of (a′) plus parallelism theory. (Parallelism theory, not direction theory.) This weaker claim is more plausible since it does not require that parallelism theory itself consist of logical truths, and I suspect that it is a more accurate reflection of Wright's real position than the claim that (a) is logically equivalent to (a′). I have ignored this in the text, to make my formulations sound more like Wright's formulations; note, however, that the crucial claim (L) below is formulated so as to be neutral between the stronger and weaker claims.

[11] Or, of the existence of lines *plus parallelism theory* – see fn. 10.

insofar as it fits into the emphasis in the last chapter of the book on Frege's logicism. (Wright doesn't claim to be a logicist about directions, but only about numbers; but I suspect that that is for the reasons noted in footnote 10.) Whether this interpretation ultimately makes sense is another matter: I don't see how the existence of objects of any sort can follow logically from the existence of objects of an entirely different sort. To put the point another way, I do not see how it can be maintained that a theory like direction theory that postulates new entities (directions) conditional on old entities (lines) is a theory that *cannot rationally be doubted* (or even, that cannot rationally be doubted given the acceptance of the old entities and of the standard theory (parallelism theory) involving them). Apparently it is a main task of Wright's book to make it understandable how such a theory can resist intelligible doubt, but when one looks closely at Wright's argument it is hard to find anything that ultimately helps one understand this.

In an important section near the end of the book (pp. 147–54), Wright seems to say that the interpretation just considered is the right interpretation of everything he has said so far, but that what he has said so far needs a slight modification in order to avoid the objection just raised. The main conclusion of this passage is that we should retreat from a version of logicism in which $(\eta^=)$ counts as a logical truth to a version in which it doesn't: as he remarks in a footnote to this passage, 'Now, in contrast [to the position earlier], we are free to avail ourselves of $\eta^=$ as a non-logical axiom' (pp. 184–5). On the face of it, this might seem to be giving up anything reasonably called logicism: isn't it the whole essence of logicism that one doesn't need *non-logical* axioms to develop number theory (and other branches of mathematics)? Wright's answer to this is that the only non-logical axiom that one does need, namely $\eta^=$, is a very special one: it constitutes an explanation of the concept of number. Similarly, I assume, $(D^=)$ constitutes an explanation of the concept of direction. Apparently, then, the main idea of (L) is to be preserved; the only modification is that 'logically equivalent' is to be replaced by 'conceptually equivalent'. Let's call this modified version (L').

As I've mentioned, Wright's main reason for moving from (L) to (L') seems to be to remove the mystery of how logic can guarantee the existence of objects. The discussion prior to the introduction of the modification in (L) (more strictly, the modification in the analogous claim about numbers) begins as follows:

The natural view of the matter is simply that there is a tacit existential *assumption* built into $\eta^=$: the assumption that where Fx is any (finitely instantiated) concept, there is such a thing as $\eta x{:}Fx$; and that this is an assumption for which no warrant has so far been provided, let alone a warrant grounded in pure logic

. . . But the Fregean view would be quite different: it would be that the very
fact that it is possible to establish a sortal notion of number by reference to
concepts of higher-order logic serves to *disclose* tacit existential commitments
in those statements whose truth we are fixing as necessary and sufficient for
that of statements of numerical identity. (In just the same way, the very fact
that it is possible to establish a sortal notion of direction by reference to lines,
and parallelism, serves to disclose tacit existential commitments to directions in,
among others, statements asserting parallelism of lines.) So there is no existential
assumption being made: if we have indeed succeeded in determining a sortal
notion of number in a satisfactory way, the existence of $\eta x:Fx$ for a given
(finitely instantiated) Fx is a matter not of assumption but of the truth of
appropriate statements in higher-order logic. Which, if either, view is correct?
(p. 148)

The answer to this last question is provided several pages later:

So: is it right to regard the Fregean as merely importing existential contraband
via $\eta^{=}$, or does that principle reveal, as he would prefer, a genuine existential
commitment to numbers flowing from pure logic? The answer, I think, is
neither. (p. 152)

The way between the two answers is to be provided by the view that
it isn't logic, but the explanation of our concepts, from which the
existence of entities flows. This is taken to be a significant concession,
for it implies that:

logic has no leverage to exert on someone who refuses to accept the admissibility
of anything but an "austere" [i.e., ontological reductionist] reading of
equivalences like $\eta^{=}$. (p. 152)

As I understand him, Wright is claiming that the nominalist can maintain
his or her nominalism without flouting logic; however, to do so he or
she must refuse to accept the concept of number entirely, or at least,
refuse to accept the explanation of the concept given by $\eta^{=}$.

I will soon return to the question of the import of this concession to
the nominalist; but first I want to deal with a more important issue,
namely, whether (L′) is any more believable than (L). I must confess
that I do not see how the idea that the existence of objects can flow
from an explanation of concepts is any less mysterious than the idea
that the existence of objects can flow from pure logic. The idea that the
existence of numbers flows from the explanation of the concept of
number is reminiscent of the ontological argument for the existence of
God, according to which it follows from the very concept of God that
God exists. Does Wright have anything to say which makes (L′) look
any more respectable than this analogy would suggest?

Wright's clearest attempt to explain how the existence of objects can
flow from the explanation of a concept occurs in a paragraph on

pp. 151–2. There, after having reviewed the way in which the terms 'direction', 'number' etc., are introduced – viz., by passing from a theory N containing an equivalence relation or equivalence operation to a theory A_N that results from it by abstraction as discussed in section 3 – he writes:

Now, how exactly do these reflections illuminate the sense of incoherence which the ideas that a person might lack an age, a geometric figure a shape, a finitely instantiated concept a number of instances etc. are apt to inspire? Like this, I think. When the concepts in question are explained as they are explained: when that is, the obtaining of an equivalence relation between a pair of things of some other sort is stipulated as necessary and sufficient for the truth of identity-statements under these concepts, then it is a straightforward consequence of the explanation, coupled with the reflexivity of the relation of the *definiens*, that, for example, any particular finitely instantiated concept has a number. . . . The suggestion that a finitely instantiated concept might *lack* a number, or any of the other corresponding suggestions, is thus in conflict with the form of explanation which abstract sortal concepts of this family receive, together with the reflexivity of the relations of the relevant *definientia*. Hence the sense of incoherence which such suggestions inspire in us: for the suggestions sound like expressions of scepticism; and no such scepticism requiring recourse to the relevant concepts for its very formulation can be anything but incoherent. (pp. 151–2)

It seems to me quite reasonable to say that the way to explain the concept of natural number or cardinal number is to formulate the numerical theory A_N; this theory is so bound up with the application of the notion of number as to be reasonably regarded as essential to that notion. In the same way, it seems quite reasonable to say that the way to explain the concept of God is to formulate the theory that God is an omniscient, omnipotent and perfectly benevolent being that created the material universe. It follows from the concept-introducing theory A_N that numbers exist (since A_N entails each atomic sentence of the form $\eta_x(Fx) = \eta_x(Fx)$, which leads to $\exists y(y = \eta_x(Fx))$ by existential generalization); it follows even more directly from the concept-introducing theory of God that God exists, since 'God is omnipotent' is already an atomic sentence on which one can existentially generalize. To say this isn't to say in any interesting sense that the existence of God flows out of the concept of God: for the concept-introducing God-theory needn't be true. Similarly, the numerical theory that serves to introduce the concept of number needn't be true, or at least, we have as yet no argument that it need be true; perhaps there is an independent argument for the conclusion that it must be true, but my point is that such an independent argument is required.

It is useful to put the point in another way. The fact that the concept-introducing theory of God needn't be true entails that 'God is omnipotent, omniscient etc.' is not a conceptual truth; what is a conceptual truth is only '*If God exists then* God is omnipotent etc.'. Similarly, though the theory used to introduce the notion of number includes $(\eta^=)$, this doesn't make $(\eta^=)$ a conceptual truth; the conceptual truth rather is

> $(\eta^{=*})$ *If numbers exist* then
> $[\eta_x(A(x))=\eta_x(B(x)) \equiv$ there are exactly as many x such that $A(x)$ as x such that $B(x)$.]

And $(\eta^{=*})$ unlike $(\eta^=)$ does not entail the existence of numbers. The last sentence of the paragraph last quoted seems to suggest that there is something incoherent in accepting the concept of number but refusing to accept the passage from $(\eta^{=*})$ to $(\eta^=)$, but I cannot see that there is any more plausibility in this than in the analogous assertion about God.

At the end of section 2, I said that there was little in Wright's book that could be construed as an argument for claim (S), the claim that separated the Strong Priority Thesis from the Weak Priority Thesis, and hence the claim that was crucial in Wright's case for platonism. The only thing that I have been able to find that can be construed as an argument for this claim is the argument that I have just considered; I conclude, then, that Wright's positive argument for platonism is unsuccessful.

Perhaps, however, Wright's book is more successful if viewed not as a positive argument for platonism, but as a defence of platonism against epistemological objections. Recall that it is after all a main goal of the book to argue that the problem that Gödel's (pseudo-explanatory) appeal to a faculty of 'mathematical perception' was supposed to solve is not a real problem. Recall also the passage on p. 152 quoted above – the passage where Wright granted that nominalism did not flout logic, where he weakened his claim to the claim that nominalism flouts the logicist's concept of number. This passage is most naturally interpreted as abandoning the idea of a positive argument for platonism, and resting content with showing that we can accept the platonist's concept of number and hence (Wright thinks) the ontological commitment to numbers without requiring a special epistemology such as Gödel's (or alternatively, Quine's and Putnam's) to support it.

Unfortunately, I do not think that even this claim has been given support. Wright's position, evidently, is roughly this: our ability to know pure mathematical truths, such as '$1+3=4$', ultimately derives from our ability to know claims like (iii), which in turn derives from (or is equivalent to) our ability to know claims like (iii'). In this section, however, I have argued that Wright's attempt to get knowledge of (iii) from knowledge of (iii') must rely on the idea that the existence of numbers flows from the very concept of number, and that this claim is very hard to support. Perhaps it would be a small move in the right direction for a defender of the idea that *God's* existence follow from the very concept of God to admit that this didn't provide a positive argument for the existence of God (since the atheist could reject the concept of God) but merely served to provide a satisfactory epistemology for theists and hence to remove the need for an alternative positive argument for God's existence. But this certainly would not be a *large* step in the right direction, and nothing that I can find in Wright's discussion of numbers seems substantially disanalogous.

5

I am afraid that the reader of this review may get the impression that I do not take Wright's book very seriously. This would be far from the truth. The attitude I have toward it is rather like the attitude I have always had about Carnap's tantalizing paper 'Empiricism, semantics, and ontology' (a paper whose doctrine has considerable affinities to Wright's): as one reads either Carnap or Wright, there seems to be something deeply and importantly correct about what one is reading, but it all seems to vanish when one tries to get clear just what it is.[12] It has not been my goal to dismiss Wright's views; rather, the review is intended in large part as a challenge to Wright or someone else to explain why the skeptical reaction I have been expressing is misplaced.

[12] Supplement A to Carnap's *Meaning and Necessity*.

6

Can we dispense with space-time?[1]

1 Relationalism versus Substantivalism

According to the relational theory of space-time, the physical world contains spatio-temporal aggregates of matter (spatio-temporally extended physical objects, spatio-temporal parts of such objects and aggregates consisting of spatio-temporal parts of different objects); these aggregates of matter are interrelated in various ways by various geometric (and also non-geometric) relations, but the physical world does not contain a space-time over and above these aggregates of matter and their interrelations.

It is tempting to put this doctrine by saying that there are no space-time regions, but only aggregates of matter. This formulation might be faulted, for a relationalist might want to 'logically construct' regions out of aggregates of matter, and given such a 'logical construction' the relationalist will assert that regions do exist. But a precondition of finding such a method of 'logically constructing' regions out of aggregates of matter is that we be able to find a satisfactory theory that does without any talk of regions and talks of aggregates of matter alone. Thus the tempting formulation, while it might be strictly inaccurate, is correct in spirit.

According to the substantival view of space-time, the physical world contains not only aggregates of matter (physical objects, their spatio-temporal parts etc.), but also (over and above these, i.e., not logically

[1] I talked about the topics of this paper in a seminar at MIT in the spring of 1984. I learned a great deal as a result of the sustained discussion by several participants in that seminar, especially Ned Block and Paul Horwich; as a result, this paper includes discussions of many possible strategies for the relationalist that I would otherwise have overlooked. I'd also like to thank Paul Teller for helpful comments on an earlier draft.

constructed from them) space-time and its spatio-temporal parts. A 'part of space-time' is of course just a space-time region.[2]

A common element to relationalist views and substantival views is the use of the geometric notion 'is a part of'. From a substantival viewpoint, it is natural to regard this notion as one that applies fundamentally to space-time regions (though of course it applies derivatively to the objects or other aggregates of matter that occupy those regions). From this viewpoint, assumptions that one makes about the part-of relation are assumptions about the structure of physical space-time. A theory of space-time structure, then, will postulate a large bunch of objects called regions; and among the predicates that it will use to talk of regions will be a 2-place predicate '$\subseteq$' meaning 'is included in' or 'is a part of'. For classical physics anyway, this will be assumed to obey the postulates of *mereological algebra* – i.e., the postulates of Boolean algebra, except modified slightly so as to deny rather than assert the existence of a smallest element (a region that is included in every region – an 'empty region', so to speak). In addition, postulates are needed to guarantee the existence of sufficiently many regions. (Usually this is done in part by a completeness schema that tells you, for any condition on regions expressible in the language, that if there are regions satisfying that condition then there is a smallest region in which all regions satisfying that condition are contained.)[3]

Indeed we normally assume a further mereological postulate, called atomicity, in developing classical physics; it says that every region includes *minimal* regions, i.e., regions with no proper parts. Such

[2] From a substantivalist viewpoint it is rather natural to identify a physical object or other aggregate of matter with the part of space-time that it occupies. If this identification is made, then the physical world consists *entirely* of space-time and its spatio-temporal parts; the distinction between space-time regions 'occupied by physical objects' and those not so occupied is explained in terms of the properties of the regions in question (and perhaps of surrounding regions). I doubt that there is much philosophical interest in the distinction between substantival views that accept the identification and those that reject it, and so I will remain neutral on whether a substantivalist would postulate aggregates of matter over and above regions of space-time.

[3] The strength of the schema depends on the background language and background theory. If the background language and theory is rich enough (e.g., if set theory is part of the background), then the schema will assert the existence of more regions than we ever need assume in practice, and so we might contemplate weakening it. Indeed, in the case of rich background theories like set theory, if we don't weaken the schema slightly, or restrict some further geometric assertions (such as Dedekind continuity) which in their natural unrestricted form are assertions about *all* regions, then we may get unwarranted empirical consequences. See sections 5 and 6 of essay 4. (Also, on a related matter, see the remark about non-measurable regions in the third paragraph of fn. 24 of the present essay.)

minimal regions are called 'space-time points'. (Normally, the choice of primitives in our geometry and in the rest of our physical theory is heavily dependent upon this atomicity assumption, though it may well be that an alternative choice that does not presuppose atomicity is possible.)[4] When the atomicity postulate is assumed in addition to the postulates of mereological algebra, then we can if we like think of regions as non-empty sets of physical points. (*Arbitrary* non-empty sets of physical points, if we have the completeness schema in a language that contains set theory.). However, *this set-theoretic representation of regions depends on these strong structural postulates.* I don't think that any of these postulates (atomicity, completeness, or the ones embodied in the postulates of mereological algebra) should be treated as 'necessities of thought'; they are just as much subject to rational revision as are the other geometric assumptions about space-time structure, though perhaps their empirical content is less direct and more dependent on the rest of physical theory than is the case with some of the other geometric assumptions.[5] *So the possibility of representating regions as sets of points depends on geometric assumptions about the structure of space-time which we might well come to revise.*[6]

From a relationalist point of view, of course, the theory of the part-of relation is not part of the theory of space-time structure (since from the relationalist viewpoint there is no such thing as space-time structure); but there too 'is a part of' seems like a geometric notion, and the assumptions involving it are no less empirical than are other geometric assumptions. And indeed it is usual to make the same assumptions involving the part-of relation that one makes on the substantival version: aggregates of matter, like regions of space-time, are assumed to obey the laws of mereological algebra, and usually the atomicity and completeness postulates as well. Because of this common theory of the part-of relation, it is possible to think of 'the relationalist's universe' as

---

[4] For instance, the usual betweenness relation of geometry might be abandoned in favour of a convexity predicate; predicates comparing scalar field values at different points might be abandoned in favour of predicates comparing integrals of associated scalar densities; etc. It isn't clear how to axiomatize our theories using such an alternative choice of primitives, but I know of no obvious reason to suppose that there is no reasonable way to do it.

[5] For various reasons it seems likely to me that any serious revision of the laws of mereological algebra (e.g., the distributive law) would be accompanied by an abandonment of the atomicity condition.

[6] If what I have just said about the empirical nature of mereology is right, it is incorrect to speak of the 'logic' of the part-of relation, as I did in Field 1980. (Another reason why this is incorrect is given in part 4 of essay 4.) I will make some comments on the significance of this in fn. 27.

a proper part of 'the substantivalist's universe'[7]: on this reading, the relationalist's 'aggregates of matter' are just space-time regions, but they are very special regions, namely, regions fully occupied by matter. From this viewpoint, it is the acceptance of *unoccupied* space-time regions (or more accurately, space-time regions that are *not fully occupied*) that distinguishes the substantivalist from the relationalist.[8]

Notice that it is quite misleading to describe the relationalist/ substantivalist dispute by saying that whereas the relationalist believes in aggregates of matter, the substantivalist believes also in space-time points. For in the first place, the substantivalist may not believe in space-time points: he or she may not believe that the process of taking smaller and smaller space-time regions has a limit, i.e., that there are regions with no proper parts. The assumption of minimal parts is a considerable technical convenience, but it may well be avoidable. In the second place, the relationalist may well want to employ an analogous technical convenience: he or she may want to hold that the process of taking smaller and smaller spatio-temporal aggregates of matter has a limit, i.e., that there are instantaneous point-particles with no temporal extent and no spatial volume. (The term 'point-particle' is not intended to imply that they have non-zero finite mass; minimal aggregates with infinitesimal mass, or with zero mass but non-zero mass density, are 'point-particles' in the sense intended here.) I believe that the introduction of such minimal elements is at least as much of a technical convenience to the relationalist as to the substantivalist, and that if a relational theory with minimal elements is feasible, there will be at least as much difficulty in reformulating it without the technical convenience of minimal elements as there is in reformulating the substantival theory in this way.

So we ought to keep distinct issues distinct. There is on the one hand the question of minimal elements, which arises on both substantival and relational views. And there is on the other hand the basic dispute between substantivalism and relationalism, viz., whether we should accept not fully occupied space-time *regions*.

In what follows my concern will be with the latter issue alone. Because it is at least technically convenient to do so, my focus when I consider substantival theories will be on theories that postulate space-time *points*, and my focus when I consider relational theories will be with theories that postulate *instantaneous point-particles* in the sense just considered. But I think it will be clear that the same issues would arise for the relationalist if minimal elements were disallowed in both theories.

---

[7] Assuming the substantivalist accepts the identification suggested in fn. 2.

[8] Again, the relationalist may accept not fully occupied regions *as logical constructions*; but this presupposes a more basic theory that makes do without them.

2 Monadicism: A Third Alternative?

It seems to me that there are two main sorts of considerations that favour a substantival theory of space-time over a relational theory. The first sort of consideration, about which I will say only a little, is the sort of consideration that has led to the ascendancy of field theories over action-at-a-distance theories in physics. I will make some remarks about this in section 3. A second sort of consideration, involving what I will call *the problem of quantities*, and of which I think the standard problem about acceleration is a special case, will be the focus of the bulk of the paper (sections 4–10).

Before discussing the considerations favouring substantivalism over relationalism, however, I want to say a few words about a third viewpoint which has sometimes been proposed as a way between the other two doctrines. The first explicit statement of the view that I know of is by Paul Horwich (1978) – in conversation, Horwich calls the view 'monadicism'.[9] Monadicism, like relationalism, denies the existence of space-time (except as a logical construction). But unlike relationalism, it does not try to make do with aggregates of matter and relations among them; for in addition to relations, it allows *primitive monadic properties of spatio-temporal location.* Consider a substantivalist view according to which there are space-time points (i.e., minimal regions). Such a view presumably recognizes a property of *occupying point p*, for each space-time point p.[10] On the substantival view, such monadic location properties are not primitive: they are obtained from the dyadic relation *occupies* by instantiating one argument with a space-time point. The monadicist can not accept this, since he or she rejects space-time. But instead of following the relationalist in regarding such properties as illicit (or in trying to reconstruct them in terms of relations among objects), the monadicist accepts them as primitive properties that a piece of matter can meaningfully be said to have or to fail to have.

It seems to me that the appeal of monadicism as a third alternative beyond substantivalism and relationalism rests largely on an equivocation between what I will call the *predicate interpretation* of it and the *higher order interpretation.* Generally, when one speaks of reducing one's ontology by expanding one's stock of primitive *properties*, what one

---

[9] The view has more recently been advocated by Paul Teller (forthcoming). Views somewhat similar in spirit have been advocated by Larry Sklar (1974) and Brent Mundy (1983); these will be discussed later.

[10] If it does not identify a piece of matter with the region that it occupies (cf. fn. 2) but instead contains a primitive notion of a piece of matter *occupying* a region of space-time.

really means is reducing one's ontology by expanding one's stock of primitive *predicates*.[11] Such a reduction can often be important: it is often an important gain in the overall simplicity of a theory to reduce its ontology by expanding its ideology. There are many signs in Horwich's paper (1978) that this 'predicate interpretation' of the rejection of space-time in favour of properties is what he has in mind. For instance:

Each [substantivalism and monadicism] recognizes a set of properties expressed by predicates of the form 'occurs at time t'. *The difference is merely that according to [substantivalism], these are two place predicates, similar to "is the brother of John", in which one place has been filled by a name or description of an instant. Whereas on [the monadicist position] they are pure monadic predicates* [my italics]. (p. 411)

Unfortunately, when one looks at the way in which properties are used in place of space-time points on the monadicist proposal, *predicate* monadicism does not look at all attractive, nor is it apparent how the details of it could be carried out. The reason it is unclear how it would be carried out is that it is unclear what the monadicist analogue is of substantivalist assertions that quantify over space-time points or other space-time regions: the idea of just replacing talk of points or regions by talk of location properties is unavailable to the predicate monadicist, for the predicate monadicist does not quantify over location properties, but merely uses location predicates. The reason predicate monadicism looks unattractive (even if it can be carried out) is that to accept a monadicist theory would be to accept a theory with an infinite (indeed, *uncountably* infinite) stock of semantically primitive predicates – one such predicate corresponding to each of the space-time points or regions recognized by the substantivalist. It seems to me that because of this, a predicate monadicist theory is rather unappealing.[12]

What then of the higher order interpretation of the monadicist position, i.e., the interpretation of it on which one quantifies over location properties, and makes higher order predications of these properties? This *is* a workable position; however, I do not see it as significantly different from substantivalism (and indeed, it counts as a substantivalist view, as substantivalism was defined in the previous section.)

More explicitly: if one takes a substantivalist view that quantifies over aggregates of matter and space-time regions, and which does not quantify over or have names for properties of aggregates or of regions (though

---

[11] Indeed, it seems to me that the notion of 'primitiveness' makes sense only on this interpretation, though doubtless that is controversial.

[12] This dismissal is rather quick, but there will be a slightly more detailed discussion of two similar proposals in sections 7 and 8.

of course it uses predicates of aggregates and of regions), then nothing can stop us from renaming regions 'location properties'. Moreover, the only primitive predicate in the substantivalist theory that relates aggregates to regions is the predicate 'occupies', and nothing can stop us from rewriting this as 'instantiates'. (Some further renaming of predicates will likewise be appropriate: e.g., if the substantivalist theory employs the predicate of one space-time point being *between* two others, we should rewrite this as a predicate of three location properties being such that the location attributed by one is between the locations attributed by the others.) We now have a theory in which the only *individuals*, i.e., *non-properties*, are aggregates of matter. The problems that beset predicate monadicism do not arise: there is no difficulty with uncountably many location properties, because these are simply entities that we quantify over and assert things about. .

Is higher order monadicism, as just developed, compatible with substantivalism? Well, it is not compatible with what might be called *strong substantivalism*, which is substantivalism as previously defined conjoined with the further claim that space-time regions are individuals as opposed to properties; but it is compatible with (indeed, it is a version of) substantivalism without this further claim.

If for the moment we assume that one or the other of these forms of substantivalism is correct, which form should we prefer, strong substantivalism or higher order monadicism? I am tempted to say that it doesn't much matter: that what matters is whether there are space-time regions over and above aggregates of matter, not whether, if there are such regions, they are thought of as 'individuals' or 'properties'. It seems to me that there is a genuine importance in the distinction between ontology and ideology: if one can do without a category of entities (e.g., meanings) by simply invoking a predicate or two (such as synonymy and meaningfulness) instead, then other things being equal it is advantageous to do so. But it is less clear to me that it should be of much importance how one divides up the entities in one's ontology into 'individuals' and 'properties'. The apparent interest of this derives mostly, I think, from the genuine interest in the distinction of ontology from ideology; for it derives from a usage of 'property' in which to talk of a thing's properties is simply a loose way of talking about the ideology used to describe the thing.

Despite this, I think that there are real issues behind whether a strong substantivalist formulation or a higher order monadicist formulation is more natural. Suppose that, quite independent of any need we might have for space-time regions in our physical theories, we had need for physical properties; suppose indeed that we need not only physical properties (and of course an instantiation predicate that allows us to connect up these properties with the things of which they are properties), but also higher order predicates (beyond an instantiation predicate) that apply to these physical properties. If so, it would seem rather natural to introduce location properties as well, and to allow higher order predicates of them; and then, if these location properties served all the purposes of space-time regions, that would certainly motivate our speaking in higher order monadicist terms rather than in strong substantivalist terms. One has to be careful here: the requirement that location properties serve all the purposes of space-time regions is stronger than might first appear, for reasons I will comment on in the paragraph after next. But if these conditions were satisfied, I would regard higher order monadicism as preferable to strong substantivalism.[13]

In my opinion, however, these conditions are not satisfied: there is no need for a strong substantivalist to postulate properties in physical theory (let alone, to postulate properties to which he or she attaches higher order predicates beyond an instantiation predicate). If this is right, then if one turns a strong substantivalist theory into a higher order monadicist theory by calling space-time regions 'location properties', *those 'location properties' will be the only properties postulated in the theory*. I think that under these conditions, it is clear that higher order monadicism is just strong substantivalism transcribed into misleading language.

I have baldly asserted that the strong substantivalist does not need to quantify over properties. I think that this can be defended, and indeed the formulation of a strong substantivalist physical theory which I refer to later on (that given in Field 1980) is one that does not quantify over properties (even in dealing with physical fields). Nevertheless I do not want to rest upon that here. So it is worth noting that even if one *does* have reason to quantify over properties (and to introduce higher order predicates of these properties, beyond an instantiation predicate), a strong substantivalist formulation may be preferable. To see this, reflect that whatever reasons might induce one to quantify over properties of physical objects (and make predications of these properties) may well induce someone who accepts space-time regions to quantify over properties *of space-time regions* (and make predications of *these* properties).[14] In that case, if one treats space-time regions as location

<hr>

[13] 'Preferable to' probably should not be taken here as meaning 'more likely to be true', because as the previous paragraph suggests I'm not sure that there is a factual issue here. But my point is that under these conditions a higher order monadicist formulation would more naturally convey the import of the total theory.

[14] Indeed, the argument for postulating physical properties that I used to find most persuasive – Putnam's (1970) argument, that we need to quantify over properties of space-time regions in order to deal with physical fields – is an argument that would *only* require ascribing properties to space-time regions.

properties, the properties of space-time regions become properties of properties, and predicates of them become predicates of properties of properties. It is very unlikely that we need predicates of properties of properties elsewhere in physical theory; if this is right, then the invocation of space-time regions again forces us to introduce entities of a type we didn't have before, and a higher order monadicist formulation serves only to disguise this fact.

What this discussion shows is that under certain conditions, a higher order monadicist transcription of a strong substantivalist theory would be preferable (in one respect) to the strong substantivalist formulation. The conditions under which this would be so are that in the strong substantivalist formulation one invokes properties of aggregates that are of a higher type than any properties of space-time regions which one invokes. If these conditions are met, then it is natural to alter the theory formulation by construing space-time regions as properties, since doing so reduces the number of individuals without introducing properties of a type higher than one is already employing elsewhere. But if the conditions are *not* met, then the strong substantivalist formulation is preferable. In my view, these conditions will not be met. (I would defend this by arguing that there is no need to quantify over properties at all; but there are other points of view from which the same conclusion could be reached.)

It also seems to me that there is an independent consideration that slightly favours strong substantivalism over the monadicist version of substantivalism. The consideration I have in mind is based on Leibniz's indiscernibility argument. This venerable argument seems to me to have no force whatever against strong substantivalism, whereas I think that it *may* have *some* force against monadicism. The issues involved here are rather tangential to the main body of the paper, so I will confine them to a footnote.[15]

[15] In its barest bones, the Leibniz argument says:

a) that it is unreasonable to suppose that *the universe could have been just like it actually is except with everything moved over one mile in some specific direction*: a universe that was qualitatively like the real universe, in a sense of 'qualitatively like' in which a systematic shift of the positions of all parts is not sufficient for qualitative difference, would just *be* the real universe.

b) that the substantivalist has to grant what (a) says that it is unreasonable to grant.

Now, one possible reply to this argument (one that Horwich himself develops fairly persuasively in the article mentioned) is that the supposition italicized in (a) isn't unreasonable in the way that it first appears, and that this is so because the possibility granted in (a) can't be used to generate epistemological problems about which sort of

cont. over page

In any case, there is really no need for me to take a stand on the dispute between higher order monadicism and strong substantivalism: both positions agree that there are regions of space-time over and above aggregates of matter (though they disagree as to whether these regions are 'individuals' or 'properties'), and for purposes of this essay, that is all that matters.

---

universe one is in. Another possible reply is that the modal assertions in (a) are simply unintelligible. If either of these sorts of replies are fully adequate, then neither form of substantivalism (strong or monadicist) has anything to worry about in the Leibniz argument. However, for many people a disquiet remains: there seems to be a sense that one can give to the modal assertions in (a) which does make those assertions unreasonable, as (a) claims. If these people are right, then substantivalism can be made satisfactory only by showing, *contra* (b), that it does not license these modal assertions (so understood).

It seems to me that there is no obvious reason why a strong substantivalist should grant (b): he or she can hold instead that 'individuation of objects across possible worlds' is sufficiently tied to their qualitative characteristics so that if there is a unique 1–1 correspondence between the space-time of world A and the space-time of world B that preserves all properties and relations (including geometric relations among the regions, and occupancy properties like *being occupied by a round red object*), then it makes no sense to suppose that identification of space-time regions across these worlds goes via anything other than this isomorphism. (And if there is a non-unique isomorphism, it makes no sense to suppose that identification across possible worlds fails to respect at least one of the isomorphisms.) If the strong substantivalist takes this line, then he or she will not accept (b). He or she would use the same rationale for ruling out the allegation that there could be a world qualitatively just like ours, but with the feature that whereas in our world it is Electron A that has qualitative characteristics F and Electron B that has qualitative characteristics G, in the other world it is the other way around.

Indeed, as Horwich observed in the aforementioned article, there is a Leibnizean indiscernibility argument against the existence of electrons that is just like the Leibnizean argument against strong substantivalism, and the obvious unsoundness of the former makes it clear that the latter is unsound as well. My only reservation about Horwich's discussion of this matter is that he confines the focus of the doubt in each case to premise (a). My view is that there are many different senses of 'possible' and 'could', and that which of (a) or (b) one denies depends on which sense one has in mind. But in *any* sense of 'could', *one* of (a) and (b) clearly fails in the electron case; and if we use the same sense of 'could', then the same one of (a) and (b) fails in the space-time case as well.

Can a monadicist cast doubt on (b) in the same manner as the strong substantivalist can? Yes, but only by rejecting what seems a plausible principle: that the only predicates in a scientific language which are *not* 'qualitative' are those which contain an essential reference to not-qualitatively-described individuals in their definitions. For in order to use the point of view about crossworld identity just outlined to get the result that any world obtained from the real world in the manner described in (a) is just the real world all over again, one needs the assumption that predicates like 'occupying point p' are non-qualitative (even though predicates like 'occupying the point which is the centre of mass of the universe' would be qualitative). For a strong substantivalist, the explanation of their not being qualitative is clear: they refer to a specific individual in a manner other than by description. For the monadicist, however, there are no individuals at issue here: *being at point p* is just a primitive property. The explanation of why it shouldn't count as 'qualitative' is then quite unobvious.

3 Fields

I now turn to the first of two considerations that seem to me to favour a substantival theory of space-time over a relational theory: namely, the sort of consideration that has led to the ascendancy of field theories over action-at-a-distance theories in physics. I won't say *much* about this, but it is worth saying a bit, because some philosophers have seen the relevance of field theories to the substantivalism/relationalism dispute differently.

What is a field theory? As I see it, a field theory is simply a theory that assigns causal properties to space-time points or other space-time regions directly (as opposed to indirectly, via matter that occupies those points or regions). (Or to be more accurate, it is a theory that *employs causal predicates* that apply directly to space-time points or regions.)[16] For instance, in electromagnetic field theory we assign to each point in space-time an electromagnetic intensity,[17] irrespective of whether this point is occupied by matter. Obviously this presupposes a substantival view: on a relational view, there are no points or other regions of unoccupied space-time, so the assignment of a property to such a point or region makes no sense. Consequently, it seems to me that for a physical theory to accord with anything reasonably called relationalism, that physical theory can not be a field theory. Instead of predicting and explaining the behaviour of matter in terms of fields, i.e., properties of (unoccupied as well as occupied) regions of space-time, a relationalist physical theory would have to predict and explain the behaviour of matter in terms only of that matter and other matter (e.g., the matter that a substantivalist might intuitively think of as 'generating' the relevant

[16] Indeed, nearly all talk of properties and relations in this essay is to be understood as a convenient manner of speaking, to be replaced by talk of predicates when one is being strictly accurate. (The only exceptions occur in discussions of doctrines that I do not myself advocate: they occur either in the discussion of higher order monadicism in section 2, or in the discussion of a related doctrine in fn. 44, or in the discussion of a modal defence of relationalism in sections 9 and 10. These exceptions will all be explicitly noted at the time.) That all my talk of properties and relations can be so construed is important, for it guarantees that there cannot be the sort of clouding of the issues between substantival and non-substantival views that we saw in the case of higher order monadicism.

[17] Or better, we assign various *intensity relations* to pairs, triples etc. of points. (The details of what intensity relations one uses in a treatment of fields depend on what type of field (scalar, vector, tensor) is at issue; cf. Field 1980, chs 7 and 8). One of the reasons that this treatment is better than an assignment of a numerical intensity is that it does not involve an arbitrary choice of numerical scale for the field intensities.

Of course, talk here of 'assigning intensity relations' is to be construed (in accordance with fn. 16) as a loose way of talking about employing multi-place intensity predicates. (This does not lead to the problems that beset predicate monadicism, as can be seen by consulting the details of Field 1980, chs 7 and 8.)

aspects of the field). A physical theory which is relationalist in this sense is called an action-at-a-distance theory.

As I understand relationalism, then, it is committed to the programme of replacing field theories by action-at-a-distance theories throughout physics: the electromagnetic field, the metric or gravitational fields characteristic of general relativity and so forth must all be shown to be dispensable constructs.[18] It seems to me that the question of whether it is possible to find an interesting way to dispense with fields is quite a fascinating one. The difficulties in doing this are quite considerable, but I certainly would not want to dismiss out of hand the possibility that it can be done. To my mind, the commitment to doing it is perhaps the most interesting consequence of the relationalist position. This however is not a matter that I can pursue in this paper.

Some philosophers have argued that the relevance of field theories to the substantivalist/relationalist dispute is precisely the opposite of what I have said: according to them, the rise of field theories makes it *easy* to maintain a relational view.[19] For, they say, a field theory postulates a giant object (the field) with a part corresponding to each space-time region; so whenever the substantivalist wants to make an assertion about space-time regions, the relationalist can make an assertion that makes reference only to the corresponding parts of the field. (A variant of this position is that 'the field' exists not quite everywhere, but at all points where the field values are non-zero. In virtue of the infinite ranges of the forces in physics, there will only be a very few isolated points where the field values for a given field are precisely zero[20] (and still fewer where the field values of *all* fields are precisely zero); so this is not a very substantial modification of the previous position, and it is hard to believe that a relationalist who maintains it will have any general difficulty in reformulating the substantivalist's assertions.)

I have no doubt that dispensing with space-time regions in favour of 'parts of a field' is possible, so if one wants to call this a version of relationalism then relationalism is certainly a manageable position. (It is

[18] In the case of general relativity, showing this presumably would involve splitting up the usual metric tensor into a component that is independent of the matter distribution and another component that represents gravitation. The former would be treated not as a field, but in the way that metrics are treated in Newtonian mechanics or special relativity; the problem would be to find an action-at-a-distance replacement for the second component. Many people would regard the need to split up the usual metric into two components in this way an unattractive feature of the action-at-a-distance approach, even if one assumes (what is highly unobvious!) that the action at a distance replacement of the second component could be given in a reasonably attractive way.

[19] See for instance Malament 1982, fn. 11 and Putnam 1981, p. 73, for remarks that imply this.

[20] Indeed, Newtonian gravitational theory can be presented as a field theory, and if it is there will be no such points.

manageable even for theories like Newtonian gravitational theory in which action-at-a-distance formulations are standard; for those theories can of course be represented as field theories (Poisson's equation), and we could then maintain relationalism in Newtonian physics by replacing all references to space-time regions by references to parts of the gravitational field.) It seems to me however that this 'saves' relationalism only by trivializing it: on an interesting version of relationalism, fields are just as much a fiction as is space-time.

Indeed, the natural way for a substantivalist to view a field is not as some giant physical object which occupies all (or virtually all) of space-time, but as not an *entity* at all. As I said before, a field theory is simply a theory that assigns causal properties to space-time regions directly ('directly' as opposed to 'via objects that occupy those regions'). To put it in a less ontologically inflationary way, it is simply a theory that employs causal predicates that apply directly to space-time regions. This means that acceptance of a field theory is not acceptance of any extra *ontology* beyond space-time and ordinary matter, it is merely acceptance of an added sort of ideology. From this 'fields as ideology' viewpoint, the method of 'saving' relationalism mentioned above makes no sense. In order to 'save' relationalism in the way mentioned, we first have to alter the way we think about field theories: we have to view them as postulating entities whose geometric properties are exactly the same as the geometric properties that the substantivalist ascribes to space-time. (Or, *almost* exactly the same, in case the relationalist views 'the field as existing' only where its values are non-zero.) Small wonder that after giving field theories a strained interpretation according to which fields are entities that have the structure of space-time built into them, a separate space-time is then dispensable.[21]

---

[21] It might be objected against what I have said that a substantivalist can not easily give a non-arbitrary criterion for distinguishing between on the one hand matter and on the other hand empty parts of space-time where the field values are non-zero. *Having mass-energy* is not a distinguishing characteristic, for on most field theories space-time can have mass-energy ('the mass-energy of the field') even where there is no matter in any ordinary sense. If so, the distinction between those points where there is mass-energy and those where there isn't is a more natural distinction from a substantival point of view than the distinction between those points where there is matter and those where there isn't. So isn't it odd to use the latter distinction rather than the former in setting the relationalist programme?

Even if we grant that the matter/non-matter distinction is hard for a substantivalist to draw, I don't see how anything follows for relationalism. The relationalist has no analogous difficulties: for the relationalist, talk of 'the mass-energy of the field' is a fiction used to explain the behaviour of matter. If the relationalist can give a theory that explains the behaviour of (what he or she counts as) matter that does not utilize this fiction, why should he or she care if the substantivalist has a hard time drawing a line between what the relationalist counts as real and what the relationalist counts as a fiction?

As I have said, I think that by attributing to the relationalist the view that field theories must be replaced by action-at-a-distance theories, I am attributing an interesting though quite difficult programme. Just how serious the difficulties in carrying out this programme are is not something I will discuss in this essay. What I want to discuss now is whether there are any *additional* difficulties for the relationalist *besides* the commitment to action-at-a-distance theories; and to do that I shall concentrate on a physical theory where an action-at-a-distance formulation is unproblematic.

4 The Second Problem for Relationalism; Outline of Rest of Paper

The most famous argument for the substantivalist view of space-time is based on Newton's bucket argument. What Newton's bucket argument shows most directly is not that we must adhere to a substantivalist view of space-time, but that we need a notion of absolute acceleration.[22] But from this preliminary conclusion, Newton drew two other conclusions. First, but of no present relevance, he concluded that we need a notion of sameness of place across time, so that absolute acceleration could be defined in the most straightforward way; this conclusion is now generally agreed not to be a good one, for there are alternative ways to define absolute acceleration not requiring a notion of inter-temporal sameness of place. Second, and what is of relevance here, he concluded that we need a substantival view of space-time (which, given his previous conclusion, amounted to a substantival view of *space* persisting through time). The reason for this second conclusion is that both the straightforward way of defining absolute acceleration (in terms of absolute sameness of place) and the alternative ways just alluded to depend on a substantival view of space-time. Given the bucket argument, we need to give some sort of account of absolute acceleration; and the problem of doing this without a substantival view of space-time (which I will call *the acceleration problem*) seemed insoluble to Newton and to many others since.

Is the acceleration problem really insoluble for the relationalist? I have no simple answer to this question. The complicated answer will be this:

1 The acceleration problem is simply a special case of a more general problem that I will call *the problem of quantities*. The problem of

---

[22] When I say 'shows most directly', I do not mean to suggest that the conclusion could not under any circumstances be controverted. Mach's programme, were it to be successfully carried out, would undermine the conclusion.

quantities arises for the relationalist in many different forms. In some of these forms, the technical details are substantially simpler than they are in the case of the acceleration problem. Indeed, a simple form of the problem arises even when motion is not considered: the problem of quantities poses a difficulty even for a relationalist description of the relative positions of objects at a given time. The acceleration problem raises no difficulties *over and above* the sort of difficulties raised by this simple case of the problem of quantities.

2   The problem of quantities could be trivially solved by a relationalist (or at least, an anti-substantivalist) willing to accept what I will call a *heavy duty platonist* position. (I substitute 'anti-substantivalist' for 'relationalist' because heavy duty platonism is not consistent with the letter of relationalism as previously defined – nor even, I would argue, with its spirit.) I will save the definition of heavy duty platonism for the next section of the essay, but for now let me merely say that it is a much more radical position than mere platonism, i.e., the mere acceptance of mathematical objects; and though I am sure that there are platonists who would not balk at this more radical position, I suspect that a great many platonists would find it unattractive. Moreover, I think that physical theories that do not rely on heavy duty platonism are far more explanatory than theories that do rely on it – if a theory that appears to rely on heavy duty platonism works, its working looks like magic unless it can be shown that the appearance of relying on heavy duty platonism is illusory. For these reasons, I do not think that adopting a heavy duty platonist position is a very interesting way for an anti-substantivalist to solve the problem of quantities.

3   There are nevertheless interesting strategies that a relationalist might propose for solving the problem of quantities, and hence (given 1) the acceleration problem. The most *prima facie* attractive strategies for solving it of which I am aware involve invoking the notion of possibility, and in the last two sections of the essay (sections 9 and 10) I will discuss the prospects of using the notion of possibility to resolve the problem of quantities. My tentative conclusion, however, is that the idea of solving the problem by the introduction of modality is not workable, and therefore that there is no attractive version of relationalism even if action-at-a-distance theories can replace field theories everywhere.

One aspect of this summary is worth emphasis: I do *not* think that the problem of quantities *must* be regarded as an insurmountable problem for the relationalist (in the way that the relationalist's need to eliminate fields may well pose an insurmountable problem); for one way to surmount the problem of quantities is simply to swallow heavy duty platonism. Although I will say a few things to try to make swallowing heavy duty platonism (and various other relationalist options) look unattractive, my main interest is simply in ascertaining what the relationalist's options are.

In particular, a main motive in writing the paper was to answer the question to be discussed in sections 9 and 10, the question of whether the invocation of modality helps the relationalist in dealing with the problem of quantities. Many people seem to think that there is a general trade-off between modality and ontology, in that whenever one apparently needs to enrich one's ontology, one can avoid such enrichment simply by invoking modal claims instead. The discussion of modal relationalism in the last two sections of the essay is designed to cast doubt on this claim. But before getting into that, it is essential to get clear on how the problem of quantities looks if modality is avoided.

5 Moderate Platonism and Heavy Duty Platonism

A. The Contrast Explained

First, let me explain what I mean by heavy duty platonism. *Platonism* – which includes both what I will call heavy duty platonism and what I will call moderate platonism – is the position that there are abstract objects such as numbers. If one is a platonist, then presumably one believes that there are *relations of physical magnitude* that relate physical things and numbers:[23] for instance, the mass in kilograms relation which might hold between a given physical object and the real number 15.3, or the distance in feet relation which might hold between two given objects and the number 7.4. The difference between moderate platonism and heavy duty platonism lies in their attitudes to such relations. According to moderate platonism, *such magnitude relations between physical things and numbers are conventional relations that are derivative from more basic relations that hold among physical things alone.* The heavy duty platonist rejects this, taking the relation between physical things and numbers to be a brute fact, not explainable in other terms. (If one likes to flaunt one's heavy duty metaphysics, one can say that there is a mysterious relation of platonic participation between physical things and numbers. But the position is the same whether or not one flaunts it.)

---

[23] That is, presumably one is willing to employ relational predicates that one assumes to be instantiated by pairs of physical things and numbers – see fn. 16.

Let us see how the moderate platonist works out his or her position that relations between physical things and numbers are derivative from more basic relations between physical things alone. Take mass as an example. One example of a mass relation that holds among aggregates of matter (or regions of space) is the 2-place relation S that holds between x and y when the mass of x is the same as the mass of y. I've explained this relation S in terms of numerical mass, but we can take it as a primitive relation. That is, we can lay down various laws for it (together with the part-of relation) that don't make any reference to numbers (e.g., that if xSy and ySz then xSz; or that given any x and y, either there is a z such that $z \subseteq x$ and zSy, or there is a w such that $w \subseteq y$ and xSw);[24] and given a sufficient body L of such laws,[25] we can prove (from these laws plus standard mathematics) two crucial theorems. First, we can prove a *representation theorem*: this says that there is at least one function m from aggregates of matter (or regions of space) into non-negative real numbers such that:

> (*) for all aggregates (or regions) x and y, $m(x) = m(y)$ if and only if xSy

[24] Note that this last law is reasonable, since we are taking parts of aggregates to themselves be aggregates; this is one reason why it is important for a relationalist to think in terms of aggregates rather than in terms of 'whole objects'.

It should be remarked however that this law would not be warranted if we allowed for the possibility of instantaneous point-masses (minimal aggregates) with non-zero and non-infinitesimal mass. If that were to be allowed, one would have to use a more complicated primitive than S in developing a representation theorem.

A final qualification is that the law as stated rules out the existence of *non-measurable* aggregates or regions, that is, aggregates or regions so bizarrely shaped that there is no assigning them a numerical mass that obeys an additivity law (even finite additivity). If we assume the full completeness schema for aggregates or regions, plus a rich set theory that includes the axiom of choice, the existence of non-measurable aggregates or regions follows from natural assumptions; so if we are operating with such a rich set theory, we should either restrict the comprehension schema for regions somewhat, or we should impose certain restrictions on the range of the quantifiers in the law to which this note is attached. Either way, a representation function (for the regions in the range of the quantifiers) is still forthcoming.

[25] One of the required laws is an Archimedean axiom; expressing it requires either some set theory or some logical devices (e.g., a second order quantifier or an ancestral operator) that are not available in first order logic. If one wants to make do without set theory and without such devices, one must use the more complicated primitive alluded to in fn. 24; and the reference to the real numbers in the representation theorem must be replaced by reference to some real-closed field or other that is very much like the real numbers. (If the primitive is chosen properly, one can guarantee that the real-closed field in question will be the same as the real-closed field that appears in the representation theorem for the first order non-set-theoretic version of the geometry.)

and

> (**) for all aggregates (or regions) x and y such that x and y are
> disjoint, $m(x \cup y) = m(x) + m(y)$.

Second, we can prove a *uniqueness theorem*: this says that if m and m^+
are functions from aggregates of matter (or regions of space) to non-
negative real numbers, and if m satisfies (*) and (**), then m^+ satisfies
(*) and (**) if and only if it is obtainable from m by a ratio transformation
(i.e., by multiplication of all values by a positive constant). As a
corollary to these two theorems, we have that if the laws L governing
the relation S hold then there is a unique function m mapping aggregates
of matter (or regions of space) into real numbers that meets the following
conditions:

> (i) $m(x) = m(y)$ if and only if xSy
> (ii)   if x disjoint from y then $m(x \cup y) = m(x) + m(y)$
> (iii) $m(x_0) = 1$.

Here x_0 is an aggregate or region whose mass is not zero (i.e., such
that $\exists y(y \subseteq x_0$ and not $ySx_0)$); it might for instance be an object that has
been chosen as the standard kilogram. We see then that relative to the
convention that (i), (ii) and (iii) are to hold (which is really just the
convention that we are to use the kilogram scale for mass), then the
laws L governing the relation S determine a unique numerical value for
each aggregate or region. *The relation that a given aggregate or region
y has to the number 15.3 (its mass in kg) isn't so mysterious: it is simply
the relation of being mapped into 15.3 by the unique function m that
accords with the conventions (i), (ii) and (iii) that have been laid down
to determine the mass scale.*
　What it means to reject heavy duty platonism is to hold that magnitude
relations between physical things and numbers must always be generated
from relations among physical things alone in something like the way
just illustrated. Call this the *moderateness condition*.
　It might indeed be argued that the moderateness condition should be
strengthened, if it is really to reflect the idea that magnitude relations
between physical things and numbers are conventional relations that are
derivative from more basic relations that hold among physical things
alone. For it might be argued that magnitude relations between physical
things and numbers can only be regarded as derivative in this way if:

> (a)   it is possible to state the laws of physics in such a way that they
>   do not involve magnitude relations between physical things and
>   numbers, but only involve relations between physical things alone

and

(b)   the formulation of the laws of physics in terms of magnitude
        relations between physical things and numbers can be derived
        from the more basic formulation of these laws (the formulation
        required in (a)), via the representation theorems for the magnitude
        relations.

Let us call this the *strong moderateness condition*, and let us call any
form of platonism on which it is accepted a form of *very moderate
platonism*. Very moderate platonism gives magnitude relations between
physical things and numbers an even more derivative status than is
required by the moderateness condition alone; though in fact I am
inclined to the view that this stronger condition ought to be regarded
as implicit in the idea of moderate platonism as I informally explained
it in the first paragraph of this section. But there is no need to take a
stand on the strong moderateness condition here: as we will soon see,
*the substantivalist has no difficulty meeting both moderateness conditions,
while it is extremely doubtful that the relationist can meet either one.*

B. Some Inconclusive Philosophical Discussion

I have no knockdown argument to offer in defence of the moderateness
condition; it is however my view that relationalism isn't all that
interesting a doctrine if such a condition is rejected.

Before trying to motivate that view, let me turn to another question:
does the substantivalist need to introduce mathematical entities in
physical theory at all? I believe that the answer to this is 'no'; but that
claim is stronger, and more controversial, than the claim that he or she
can adhere to even the *strong* moderateness condition. (The latter claim
is I think beyond reasonable controversy, at least for Newtonian physics:
see section 6.) The strong moderateness condition requires that we be
able to state the laws of physics without invoking magnitude relations
between physical things and numbers. But someone who believed that
might still believe that we need *sets of physical things*, or *sets of ordered
pairs of physical things*, or some such thing, in the formulation of
physical laws. And if we need even that much *directly*, then we may
indirectly need the higher reaches of set theory; for the higher reaches
of set theory have implications about how many sets of physical things
there are, and can thereby affect the strength of assertions involving sets
of physical things. (An example is obtainable from the remarks on
Dedekind continuity in footnotes 3 and 36.)

As I've said, I myself do not believe that we do need to invoke sets of physical things in our theories, as long as we are substantivalists; but my claim that we do not (in 1980) has given rise to some controversy.[26] Although I believe that the doubts are mistaken,[27] I do not want to presuppose that here. I will merely note that if I am right that a substantivalist needn't invoke mathematical entities in physical theory (and especially if I am also right in the further claim, the defence of which is begun in essay 3, that one doesn't need mathematical entities *outside of* physical theory), then this would greatly reduce the interest of any form of relationalism that required even *moderate* platonism. For whatever ontological advantages there might be in dispensing with space-time would be far outweighed by the need to introduce mathematical entities which would otherwise be unnecessary. (The philosophical problems raised by mathematical entities – e.g., the epistemological problem of explaining how we can have reliable

[26] See for instance Shapiro (1983).

[27] It might seem that such doubts are especially forceful given my concession (in section 1) that the theory of the part-of relation should not be counted as logic. For the idea that this theory *is* to be counted part of logic played an important role in the nominalization strategy of *Science without Numbers*. But two points need to be made. First, the nominalization strategy of that book was disjunctive, and although one disjunct (the 'second order' disjunct) relied on the idea that mereology should be counted as logic, the other (the 'first order' disjunct) did not. Since writing the book, I have come to see how to argue more convincingly that the first order disjunct is workable: cf. essay 4. And John Burgess has independently reached the same conclusion (1984), despite his own hostility to nominalism.

Second, George Boolos (1984) has recently shown that the second order disjunct might be available independently of the dubious idea of a 'logic' of mereology. The reason why I resorted to the idea of a 'logic of mereology' in one disjunct of the argument in my book was that I wanted the expressive power of second order logic, but thought that second order logic was not available to me since it quantified over classes. But Boolos argues that second order quantifiers should not be construed as quantifiers over classes of individuals, but rather as *plural* quantifiers with individuals themselves in their domain. (This applies only to quantifiers taking unary predicates as substituends; but they are the only ones needed on the 'second order disjunct' of my strategy in the book.) Boolos's argument for this conclusion in no way rests upon nominalistic assumptions; like Burgess, Boolos would have no wish to defend nominalism. I haven't yet decided whether I think Boolos is right, but if he is, it is of considerable relevance to the nominalization programme of my book. Not only would the second order disjunct be restored, but a more flexible version would become available than the one in the book. In the book, the fact that I interpreted the second order quantifiers as ranging over regions forced me to restrict second order quantification to what would traditionally be construed as quantification over classes of *points*. But if Boolos is right, I would also be able to have what would traditionally be construed as quantification over classes of *regions*. That wasn't needed for nominalizing classical field theories, but it might be needed for other sorts of physical theories.

information about them – seem far worse than any raised by space-time regions.)[28]

Since I do not want to presuppose my anti-platonist views here, the question arises whether the moderateness condition that I am invoking would have any appeal to platonists (especially to platonists who believe that mathematical entities are needed in developing physical theories). I think that there is clear historical evidence that the answer to this question is 'yes'. In the first place, representation theorems have been around since about the turn of the twentieth century, and a large group of researchers have committed themselves to the programme of finding a representation theorem to generate every numerical scale.[29] As far as I know, the first person to suggest that the successes of this programme might be used in an effort to purge physics of mathematical entities was me, in *Science without Numbers*, published in 1980. The eighty year gap suggests that the moderateness condition has had a considerable appeal independently of anti-platonist considerations. A second piece of historical evidence for the same conclusion is the fact that so many people have thought of the apparent need of an acceleration functor in Newtonian physics as a potential problem for the relationalist. If one accepts heavy duty platonism, it is hard to see what the problem was supposed to be: one could simply take co-ordinate functors defined on points of matter as primitive, and define an acceleration functor from them in the usual way, without ever having to appeal to unoccupied regions. (Thirdly, there is also weaker historical evidence that the *strong* moderateness condition has had an appeal even to platonists: a number of reviewers of my book have rejected its anti-platonism while endorsing its account of the application of mathematics;[30] and this seems tantamount to acceptance of the strong moderateness condition.[31]

---

[28] See essay 2, pp. 67–73. Note also that the epistemological problem mentioned in the text remains even if one concedes that if mathematical entities are indispensable then this serves to justify us in believing in them. Justified belief in a certain class of facts is one thing; showing that that belief does not give rise to problems, including the problem of explaining how we can have reliably true beliefs about those facts, is something else.

[29] For an extensive survey of work in this direction, see Krantz et al. 1971.

[30] See especially Shapiro (1984); also Friedman (1981).

[31] There may seem to be something puzzling about endorsing the use of mathematical entities in physics while accepting the moderateness or strong moderateness condition. For the only way that we could possibly need mathematical entities in physics is if we needed relations between physical things and mathematical entities – without such relations, the mathematical entities would be idle. And doesn't the utilization of such relations conflict with one or both of the moderateness conditions?

The answer to this is 'not necessarily'. Suppose for instance that we need mathematics in physics because we need to talk of sets of physical things. This does require recognition

cont. over page

But let us move beyond the sociological fact that many platonists *do* reject heavy duty platonism, and ask whether they *should*. Or, more precisely, let us ask whether relationalism would retain whatever interest it might initially appear to have, if the price for being a relationalist was to be a heavy duty platonist. I think it would not.

In the first place, relationalism as defined above isn't even consistent with heavy duty platonism. Relationalism is the doctrine that there is no substantival space-time; that we can make do without space-time in our basic theory of the world by relying instead on *relations between aggregates of matter*. But to accept heavy duty platonism is to introduce magnitude relations which relate aggregates of matter *to real numbers* (e.g., a *has-mass* relation which relates a given 15.3 kg aggregate to the number 15.3). This certainly violates relationalism as it was defined. But beyond that, it violates the whole spirit of relationalism. For the spirit of relationalism is that only quite unproblematic relations will suffice; whereas relations between physical things and numbers seem extremely problematic if not somehow demystified, and the heavy duty platonist rejects the only method I know of for demystifying them.[32]

Besides violating the definition of (and the whole spirit of) relationalism, the invocation of heavy duty platonism seems quite unattractive, and not merely for the 'metaphysical' reason just alluded to. It is unattractive, in addition, on 'explanatory' grounds: to adhere to it in one's physical theory is to rest content with a physical theory that is far less explanatory than is desirable. If in explaining the behaviour of a physical system, one formulates one's explanation in terms of relations between physical things and numbers, then the explanation is what I would call an

---

of a relation between physical things and mathematical entities – the membership relation – but this is a far cry from utilizing magnitude relations between physical things and numbers. (Even if numbers are themselves construed as sets, such a magnitude relation between a physical thing and a number is not expressible in terms of the membership relation.) For instance, the membership relation is uniform, in the sense that any 1–1 function from the set of physical things onto itself induces a 1–1 function of the whole set-theoretic universe onto itself, *in which the membership relation is preserved*. But magnitude relations would not be preserved by such a correspondence; this non-uniformity in them motivates the idea that when such relations hold they require an explanation.

[32] I do not claim here that *all* relations between physical things and mathematical entities are terribly problematic. For instance, it could be claimed that the membership relation between physical things and sets of physical things is no more problematic than the occupancy relation between physical objects and space-time regions (on versions of substantivalism that don't identify physical objects and regions). If this is correct, then *moderate* platonism does not have a problematic feature that it is sometimes thought to have. But this won't help *heavy duty* platonism, since it employs relations between physical things and numbers that seem intuitively to be of a much more problematic sort. See the discussion in fn. 31.

extrinsic one.[33] It is extrinsic because the role of the numbers is simply to serve as labels for some of the features of the physical system: there is no pretence that the properties of the numbers influence the physical system whose behaviour is being explained. (The explanation would be equally extrinsic if it referred to *non-mathematical* entities that served merely as labels: e.g., numeral-inscriptions, or names. So not all extrinsic explanations are platonistic. Conversely, not all platonistic explanations are extrinsic in the sense explained here. For instance, the role of sets in moderate platonist formulations of physical theories is not to serve as extrinsic labels.)

Now, extrinsic explanations are often quite useful. But it seems to me that whenever one has an extrinsic explanation, one wants an intrinsic explanation that underlies it: one wants to be able to explain the behaviour of the physical system *in terms of the intrinsic features of that system*, without invoking extrinsic entities (whether non-mathematical or mathematical) whose properties are irrelevant to the behaviour of the system being explained. If one cannot do this, then it seems rather like magic that the extrinsic explanation works. I believe that this 'intrinsicalist' requirement can easily be satisfied by the substantivalist. It may indeed be satisfiable by the relationalist, if one of the more interesting versions of relationalism considered in sections 9 and 10 can be made to work. But the intrinsicalist requirement can not be met by a relationalist who relies on heavy duty platonism to solve the problem of quantities.[34] This will become clearer in the following three sections, where the problem of quantities is explained.

6 The Substantivalist Solution to the Problem of Quantities

Let's return now to the substantivalism/relationalism dispute, and in particular to the problem of quantities. One way to put the problem of quantities is this: the representation theorems needed to generate many of the numerical functors needed in physics (e.g., distance, relative velocity and acceleration) don't appear to be available to the relationalist; for these theorems depend on structural regularities of space-time which are lost when one, in effect, throws away the parts of space-time which are not fully occupied by matter.[35] Consequently, a relationalist must apparently abandon even the weak moderateness condition.

[33] I got this terminology from Brian Loar. See his 1981, p. 62.

[34] Meeting it seems in fact to require the *strong* moderateness condition.

[35] This claim has been denied by Brent Mundy (1983): I will discuss this matter in fn. 42.

To this way of putting the problem of quantities, it might be objected that we don't really need numerical functors in physics. As will become clear below, I agree with this objection; indeed, its correctness follows from the possibility of a substantivalist physics meeting the strong moderateness condition. Consequently, I prefer to put the problem of quantities in a different way, in which representation theorems will come in only through the back door. Instead of focusing primarily on representation theorems, I will focus on *finding a reasonable ideology for physical theories*.

Before reformulating the problem of quantities (in the next section) by explaining why the relationalist has a problem in finding a reasonable ideology for physical theories, it will be useful to look at the ideology required for a substantivalist version of Newtonian physics. As far as Newtonian *kinematics* goes, a substantivalist needs only two primitive predicates in addition to 'is a part of': a 3-place predicate of (4-dimensional) betweenness between space-time points, and a 4-place congruence relation between space-time points. ('x is between y and z' means intuitively that x, y and z are on the same straight line, with x between y and z on that line; where 'straight line' includes not only ordinary lines in the 3-dimensional time-slices of space-time, but also other 'lines' which pass between different time-slices and which intuitively correspond to the paths of inertial co-ordinate systems.[36] Think of 'x,y congruent to z,w' as implying that x, y, z and w are all simultaneous – indeed, we would formally define 'x is simultaneous to y' as 'x,y congruent to x,y' – and as meaning intuitively that the distance from x to y is the same as that from z to w.)[37] There is a natural way to axiomatize the geometry of Newtonian space-time using only these primitives.[38] Let NST be such an axiomatization.

Lots of other comparative relations not taken as primitive in NST are nonetheless definable in it. For a very simple example, consider the relations

C_2: the distance from x to y is twice that from z to w,
C_3: the distance from x to y is three times that from z to w.

---

[36] Technically, the axioms on betweenness are those of 4-dimensional affine Euclidean geometry. Dedekind continuity is to be expressed as a single axiom, by quantifying over regions.

[37] A more general relation of congruence, in which x is required to be simultaneous to y and z to w, but no simultaneity is required between x and z, is definable from the more restricted kind. We don't want a still more general one, in which no simultaneity requirements are imposed, since that would induce a notion of sameness of place across time, and such a notion has no objective sense in Newtonian mechanics.

[38] It's easily obtainable from Field 1980, ch. 6, though that uses the more general relation of congruence mentioned in the first sentence of fn. 37.

etc. (where it is assumed in each case that x, y, z and w are all simultaneous). We can easily define these using only betweenness and congruence:

> $xyC_2zw\longleftrightarrow\exists u(u$ is a point, and u is between x and y, and $xuCuy$ and $uyCzw$),
>
> $xyC_3zw\longleftrightarrow\exists u_1\exists u_2(u_1$ and u_2 are points; u_1 is between x and y, and u_2 is between u_1 and y; and $xu_1Cu_1u_2$, $u_1u_2Cu_2y$, and u_2yCzw),

etc.; where 'C' abbreviates 'congruent to'.

Much more complicated examples than this could be given. Indeed, one that is relevant to the acceleration problem of section 4 is that we can define such comparative relations as 'the acceleration of S at p is twice the acceleration of T at q', or 'the acceleration of S at p is orthogonal to the acceleration of T at q' (where S and T are 'trajectory-like regions', i.e., regions of the sort that can be trajectories of point-particles,[39] and p and q are points on them) – these and other comparative predicates involving acceleration can be defined purely in terms of the three primitives mentioned two paragraphs back. I have given the details elsewhere.[40]

Indeed, every kinematic predicate required for Newtonian physics can be defined within NST. Of course, the usual numerical functors, like distance and acceleration, are not strictly speaking definable in NST, unless one takes NST to contain (besides what has so far been mentioned) a good bit of mathematics; and even then, most of them would only be definable if we also added to NST names for specific space-time points that could serve to define a distance scale, a time scale, a notion of intertemporal sameness of place, directions for spatial co-ordinate axes, and/or choice of a co-ordinate origin. (One of the ways in which NST is more 'intrinsic' than presentations of physics that rely on the usual numerical functors is that all of its statements are independent of scale, co-ordinate system etc.) But even though the functors would not be strictly definable within NST without these additions, they are all

[39] More precisely, regions that are connected in the space-time topology and that contain no two simultaneous points.

[40] Field 1980, ch. 8. (In the first order version of the theory presented there, I used an additional primitive beyond those mentioned here: a binary predicate that compared the cardinality of finite regions. However, John Burgess reports (1984, p. 393) that Kripke has shown that predicate to be definable from the other primitives. Also, in section 3 of the same article, Burgess develops a method for avoiding cardinality comparison of finite regions entirely. (In the case of defining the second derivative of spatial separation, avoiding cardinality comparisons is straightforward; in the case of second derivatives of non-geometric functions such as scalar fields, it requires a slight alteration in the primitive predicates that I used. See Burgess' predicate Q_x on pp. 390–91 of his article.))

generatable from NST (even without the additions) via representation theorems like those previously considered.

In any case, it really doesn't matter whether we can generate these numerical functors, as long as the invariant predicates that could be defined via the numerical functors can be defined without them. (By an invariant predicate I mean one that is meaningful independent of a choice of distance scale, time scale, notion of intertemporal sameness of place etc.) Since that condition is met in NST, then as far as Newtonian kinematics goes, *there is not only no difficulty for the substantivalist in meeting the moderateness condition, there is even no difficulty in adhering to the strong moderateness condition.*

When move beyond kinematics, the situation is rather similar. We need of course to add a few more primitive predicates. For instance, in order to be able to talk about mass we need a new predicate, say the predicate S mentioned earlier; and to develop the theory of gravitation we may need a couple of others as well. But from a small number of non-geometric predicates, and a small number of associated laws involving them, we can define all the other predicates we need, even quite complicated predicates, like predicates that compare the Laplaceans (or other derivatives) of the gravitational potential function at different points. (And again, we can get all the usual numerical functors (e.g., the gravitational potential functor and its gradient and its Laplacean) from this basis via representation theorems; though again these functors are not strictly speaking needed for the physics, since all the invariant comparative predicates definable from them are definable directly in the intrinsic theory.) *This shows that the substantivalist has no difficulty in rejecting heavy duty platonism, even if that rejection is taken to involve not only adherence to the moderateness condition but adherence to the strong moderateness condition as well.*

7 The Problem of Quantities for the Relationalist

The preceding discussion shows that the substantivalist has no difficulty at all in finding a reasonable ideology within which to develop Newtonian physics: the three geometric predicates mentioned for Newtonian kinematics, plus a few extra ones for mass etc., plus perhaps the predicate of set membership (see the beginning of section 5B and footnote 31) are enough. I now want to argue that the relationalist, by contrast, has a serious difficulty in developing Newtonian physics on the basis of this or any other reasonable ideology. This will be the reformulated version of the problem of quantities.

It might initially be thought that we could simply take over the substantivalist's definitions, except with the straightforward modification of replacing '*u* is a region' by '*u* is an aggregate of matter' and hence '*u* is a point' by '*u* is an instantaneous point-particle', i.e., '*u* is a minimal aggregate of matter.' But this fails very badly: indeed, *nearly every one of the substantivalist's definitions of predicates fails to work when this modification is made in them; and in nearly every case, there appears to be no further modification that is available to the relationalist which fares better.*

The problem arises even for very simple definitions, like those of the predicates C_2, C_3, C_4 etc. of the previous section. If, for instance, we try to define what it is for point-particles x and y to be twice as far apart as point-particles z and w, we will not succeed if we say that there is a point-particle u between x and y such that xu and uy are congruent to zw; for there need not be a point-particle midway between x and y. In the substantivalist's ontology, there is *something* midway between x and y, even if there is no *matter* there. By removing this aspect of the substantivalist's ontology, the relationalist has made it impossible to define C_2 in terms of betweenness and congruence in the way that the substantivalist does; and it is clear that (at least without introducing devices not yet under consideration, such as modal devices which will be considered at some length later) there is no alternative method of defining it in terms of betweenness and congruence that works any better.

And the same problem arises when we try to define C_3 in terms of betweenness, congruence and C_2; and again when we try to define C_4 in terms of betweenness, congruence, C_2 and C_3; etc.[41] *So even if we confine our attention to the simple infinite family of predicates C_2, C_3, C_4, etc.,* the relationalist has a problem in finding a reasonable ideology: he or she has to take each of these distinct predicates as primitive, and there are infinitely many of them.

It is clear, I hope, that having to take each of these infinitely many predicates as primitives precludes the possibility of developing a reasonable theory involving them. I do not want to strictly endorse the Davidsonian requirement (cf. Davidson 1965) that a reasonable theory formulation cannot contain infinitely many primitive predicates. An infinitude of primitive predicates can be acceptable, e.g., when one can

---

[41] One might initially think that we could handle the problem by introducing new ideology, such as a predicate $xySzwtu$ meaning that the distance between x and y is the distance between z and w plus the distance between t and u. With such a predicate, you can define xyC_2zw as $xySzwzw$. And you might momentarily think that you could go on and define xyC_3zw, xyC_4zw etc., e.g., defining xyC_3zw as $\exists u \exists v [uvC_2zw \ \& \ xySuvzw]$. But of course that fails: we can have xyC_3zw even if there are no u, v such that uvC_2zw. A similar pattern of failure holds whatever we take the ideological basis to be. (At least, it does if we keep to first order logic (and to a relationalist ontology). Some proposals for improving the situation by expanding our logic will be considered starting in section 9.)

recursively characterize them using finitely many axiom schemata. But that condition likewise fails here (as do various weakenings of it which I will not take the time to discuss): although a substantivalist can accept a schema like

$$xyC_{n+1}zw \longleftrightarrow \exists u(u \text{ is a point, and } u \text{ is between } x \text{ and } y, \text{ and}$$
$$xuC_nzw \text{ and } uy \text{ congruent to } zw),$$

a relationalist cannot accept this (or the modification obtained by replacing 'point' by 'point-particle'), for the same reasons that preclude acceptance of the substantivalist's definitions. Consequently, the relationalist can give no explanation of what these infinitely many primitive predicates have to do with each other. I have labelled them as C_2, C_3 etc., and thus suggested that they do indeed have something to do with each other. But the relationalist cannot explain in his or her theory what they have to do with each other; the notational similarity between them has to be regarded as an orthographic accident.

So far we have been dealing with the simple family of predicates C_2, C_3 etc. But this is just the beginning of the problem: at each further stage where the substantivalist introduces definitions of new 'quantitative' predicates, the relationalist apparently has to accept the new predicate as a further primitive. The relationalist will need, for instance, a whole bunch of infinite families of comparative velocity predicates, a whole bunch of infinite families of comparative acceleration predicates and so on. It seems quite clear that accepting all these different predicates as primitive precludes the possibility of having a useful theory.

That is the reformulated version of the problem of quantities. Now, in presenting the problem I have tacitly assumed that the relationalist accepts the restriction adhered to in the presentation of the substantival theory above: no numerical functors are to be utilized as primitives of the physical theory. Of course, this restriction need have no appeal to a heavy duty platonist. Moreover, though the restriction is required for the *very* moderate platonist (and for the nominalist), it isn't clearly essential that the moderate platonist agrees to it: it isn't clearly essential that he or she adheres to the *strong* moderateness condition. According to moderate platonism, any numerical functors that one employs must be generated from basic predicates that apply to physical things alone, via a representation theorem; but it is not totally obvious that there is anything wrong with first generating the numerical functors from such basic predicates, but then taking the numerical functors rather than just the basic predicates as primitive when it comes to formulating physical laws.

Could we in this way solve the problem of quantities without adhering to heavy duty platonism? No – and here is where our earlier formulation of the problem of quantities, in terms of representation theorems, comes in. For it turns out that the same sort of difficulties that prevent the relationalist from taking over the substantivalist's definitions also prevent the relationalist from accepting the substantivalist's representation theorems: as noted at the start of section 6, those theorems depend on structural regularities of substantival space-time which are lost on the relationalist surrogate for it.[42] Consequently, there is no possibility of exploiting the difference between mere moderate platonism and very moderate platonism: unless the relationalist is willing to buy into heavy duty platonism by building his or her theory on numerical functors not generated by representation theorems (and of course, if he or she is so willing then the problem of quantities really isn't too much of a problem),[43] then the problem of quantities remains serious. I don't say

[42] Mundy (1983) has proved a 'representation theorem' that does not rely on any structural regularities unavailable to the relationalist. However, Mundy's representation theorem is of no use in helping the relationalist avoid heavy duty platonism, because it is simply a device for generating certain numerical functions *from other numerical functions*. That is, Mundy takes as basic not predicates that relate point-particles or aggregates to each other but a functor k that relates point-particles *to real numbers*: it assigns to any three point-particles a real number that represents the inner product of two vectors that have one of the points as a common starting point and the other two points as endpoints. From this he shows how to generate co-ordinate systems, so we have a representation theorem of sorts; but it does not serve the purposes for which representation theorems have traditionally been wanted, for it does not proceed from a non-numerical basis. (Indeed, the numerical inner product primitive from which Mundy's construction starts isn't even scale-independent; though this feature could doubtless be improved upon.)

Mundy is apparently sensitive to the charge that a relationalist should not invoke functions from point-particles to real numbers in his basic formulation of physics; with the apparent intent of evading the charge that his treatment requires doing so, he sometimes rewrites the equation $k(p,q,r)=a$ (where a is a real number) as $k_a(p,q,r)$, so that we have an uncountably infinite set of 3-place spatial relations, one for each real number a. (See, for example, pp. 212 and 223 of his paper.) Clearly however this notational trick solves nothing. If one were really treating the k_as as just 3-place spatial relations, one would have to view the a's as unquantifiable subscripts. We would have uncountably many primitive predicates, and no serious theory involving them would be possible. Mundy of course does not treat the subscripts as unquantifiable. (He employs the notation $k(p,q,r)=a$ when he wants to quantify over the a.) But a quantifiable subscript is just a variable written in a different place, and by writing it in a different place we do not alter the fact that we have a 4-place relation one of whose terms is a real number.

[43] It isn't a problem at all for a heavy duty platonist who is willing to use functors that are not independent of scale, co-ordinate system etc. It seems to be rather widely recognized however that it is not very satisfactory to use co-ordinate-dependent notions in basic theory. The usual method for obtaining co-ordinate independence is to introduce tensors; but even this does not achieve independence of the scales of spatial and temporal separation or the scales for quantities like mass, so it is not satisfactory by itself. I think however that it would be possible to generalize the tensorial approach so as to achieve scale independence as well. Of course, from my own point of view there is no particular interest in doing this, since I reject heavy duty platonism.

at this point that it is unsolvable – we will consider some interesting attempts to solve it below – but I think that it is clear at least that it cannot be solved without introducing some device (perhaps modality) of a sort not yet taken into account.[44]

[44] I might as well note now one *uninteresting* way to solve it, a way that (like higher order monadicism) invokes the additional apparatus of quantification over properties, together with the use of higher order predicates (beyond the instantiation predicate) that apply to these properties. (Under 'properties' I include multi-place properties, i.e., relations construed non-extensionally.) Whereas monadicism can be obtained from substantivalism by renaming space-time regions as 'location properties', the view I have in mind now can be obtained from heavy duty platonist formulations of physics by renaming *real numbers* as properties of some sort.

To illustrate this, it is simplest to start from a heavy duty platonistic 'relationalist' formulation that employs only one predicate relating aggregates to mathematical objects. Such a formulation is given in Mundy (1983), on which I have commented already (fn. 42). It employs a predicate $k(p,q,r,a)$ relating three point-particles to a real number; $k(p,q,r,a)$ means intuitively that the inner product of the vector from q to p and the vector from q to r is a. Now here is a higher order variant of Mundy's theory – I'll call it HOM. It postulates an uncountable infinity of 'inner product relations'. These are to be three place relations among point-particles, and we are to intuitively think of each such relation S as corresponding to a real number a, in the sense that for any point-particles p, q and r, these particles in that order instantiate S if and only if the inner product of the vector from q to p and the vector from q to r is a. That's an intuitive explanation of what the relations in the family are, not part of the theory HOM; the theory itself, which I will now proceed to explain, will not talk about correspondences between relations and numbers. The theory works by introducing various higher order predicates of these relations: perhaps SUM, PROD and GR. Here, SUM(R,S,T) means intuitively that R, S and T are inner product relations for which the real number corresponding to R is the sum of the reals corresponding to S and to T; PROD is similar but with product for sum; and GR(R,S) similarly means that R and S are inner product relations such that the real number corresponding to R is greater than that corresponding to S. We can now lay down axioms governing these three higher order predicates, just like the usual axioms for real numbers. Next, we can introduce a 4-place instantiation predicate Δ. It is clear that for an inner product property S, the claim '$\Delta(S,p,q,r)$' corresponds closely to a claim of form '$k(p,q,r,a)$' in Mundy's theory. So we can use the results of Mundy's paper to get an axiomatization of a geometry, using the predicate Δ, that doesn't postulate any individuals other than point-particles. That's HOM.

It seems to me that just as it was hard to see that the difference between higher order monadicism and substantivalism was a very significant one (at least if the higher order monadicist postulated no properties beyond location properties), so it is hard to see that the difference between HOM and Mundy's own theory (a heavy duty platonistic 'relationalist' theory) is very significant. All we've done is rewrite the original theory, using the phrase 'inner product property' for 'real number', 'SUM of inner product properties' for 'sum of numbers', and 'inner product property S is instantiated by p, q, and r' for 'real number a stands in the inner product relation to p, q and r'. The same sort of trick can be performed with any heavy duty platonist theory.

8 The Problem of Acceleration as a Special Case of the Problem of Quantities

Returning now to the acceleration problem, I've said that a substantivalist has no difficulty in giving an account of absolute acceleration. The substantivalist's account depends heavily on the 4-dimensional betweenness relation which he or she employs: this allows the substantivalist to single out certain trajectory-like regions[45] as *inertial*. The account also depends heavily on the substantivalist's ontology, and in particular, on the existence of sufficiently many such inertial trajectory-like regions. That is, given any trajectory S and any point p on it, there is in the substantivalist's ontology an inertial trajectory-like region T through p and tangent to S; roughly speaking, what the substantivalist does is to define the acceleration of S at p as the rate of change of the spatial separation between S and T.

The most obvious reason why the relationalist has a problem in defining acceleration (or, in defining various comparative acceleration predicates) is that in the relationalist's ontology, the only trajectory-like regions there are are the actual trajectories – the trajectory-like regions that actually have a particle on them at all times. That means that there are very few inertial trajectory-like regions around: certainly not enough to take over the substantival account of acceleration (or more accurately, of the various acceleration predicates) directly and (though this takes more argument, which I shall not give here) not enough to enable us to find a satisfactory modification in the account.

In one of the most interesting recent discussions of the acceleration problem, Larry Sklar (1974) has suggested that the problem arises because the substantivalist thinks of absolute acceleration as acceleration relative to some entity, viz., an inertial frame or inertial trajectory-like region. Sklar maintains that the relationalist will indeed have a problem in giving an account of absolute acceleration if he or she thinks of it in this way too; but that there is an alternative way for the relationalist to think of it.

To maintain the relationalist doctrine of space and time in the face of the acceptance of absolute motions, what we must do is to deny that the predicate 'is absolutely accelerated' is a relational term! The expression 'A is accelerated' is incomplete. To complete it we must answer the question, 'Relative to what is A accelerated?' But the expression 'A is absolutely accelerated' is a complete assertion, as is, for example, 'A is red', or 'A is bored', and unlike 'A is north of'.

---

[45] For an explanation of this phrase, see fn. 39 or the text to which it is attached.

Absolute acceleration is not a relation of a thing to some other material object, even the 'averaged out mass of the universe'. It isn't a relation an object has to substantival space or spacetime itself, either. For these latter 'reference objects' don't exist. It simply isn't a relation at all. (p. 230)

Is Sklar's proposal workable? It is hard to a give a straightforward answer, because the proposal is ambiguous. The most natural way to take his words, I think, is as suggesting that we allow the relationalist to use a single 1-place predicate 'is absolutely accelerated' (or equivalently, 'is absolutely unaccelerated'). This is to be a non-relational predicate in that 'x is absolutely unaccelerated' is not to be defined in terms of a relation between x and something else (whether a piece of matter or of space-time); rather, it is to be either a primitive notion, or else defined in terms of a slightly stronger primitive (e.g., an analogue of the substantivalist's 4-dimensional betweenness) that is also non-relational.

If that is Sklar's proposal, then there can be little doubt that a relationalist ought to be allowed to use such a predicate – indeed, it is straightforwardly definable from the obvious relationalist analogue of the substantivalist's betweenness predicate. Unfortunately however, the use of such a predicate will not do anything to solve the acceleration problem, because the problem arises precisely because so few things in the relationalist's ontology are absolutely unaccelerated. All that the primitive acceleratedness predicate allows us to do is to single out those few trajectories (and segments of trajectories) that are unaccelerated from all the rest; how does that help us in defining numerical acceleration (or in defining such invariant acceleration predicates as 'has twice the acceleration of')? The answer is that it doesn't. *If we had sufficiently many unaccelerated trajectories (or unaccelerated segments of trajectories) at our disposal*, we could use them to define numerical acceleration (or more accurately, the various invariant acceleration predicates): we could just mimic the substantivalist's definitions, but using unaccelerated trajectories instead of inertial trajectory-like regions. But just as the absence of matter between two arbitrary points of matter precludes the relationalist from defining C_2, C_3 etc. from congruence, so the absence of sufficiently many accelerated trajectories (i.e., matter-filled inertial trajectory-like regions) precludes the relationalist from defining the needed acceleration predicates from unacceleratedness.

There are however two other ways to read Sklar's proposal. The first is to allow not merely a single non-relational predicate of unacceleratedness, but infinitely many predicates: perhaps a separate predicate 'has the acceleration $<r_1, r_2, r_3>$' for each triple of definable real numbers r_1, r_2 and r_3; or perhaps a more complicated infinite family of predicates, which would have the advantage over the family just given of being co-ordinate independent and scale independent. This proposal however is clearly not philosophically satisfactory: the infinite ideology (and the inability to state suitable laws that connect the predicates in the family without going outside the relationalist's ontology) make for a totally unwieldy theory – indeed, it hardly counts as a theory at all. (See the discussion of 'orthographic accidents' and the like, in the previous section.)

The third way to take Sklar's proposal – perhaps the most sympathetic, even if not the one that Sklar's words most naturally suggest – is as allowing the relationalist to use a primitive numerical (or vector-valued) acceleration functor (or perhaps a more complicated device which is scale independent).[46] If one has no problem in accepting heavy duty platonism, then this (at least the more complicated scale-independent version that was parenthetically imagined) is a perfectly satisfactory anti-substantivalist solution of the acceleration problem. (Of course, to a heavy duty platonist not concerned with such things as scale independence, a more obvious solution was already available: we could have defined acceleration from primitive co-ordinate functors. So perhaps this third construal isn't the most sympathetic way to take Sklar's proposal after all.)

It is clear, then, that the acceleration problem creates no difficulty for an anti-substantivalist over and above the sort of problem that is already raised by the problem of quantities in its simpler forms. And those problems are only problems if we reject heavy duty platonism. I am sure that there will be philosophers who not only find platonism unobjectionable, but who also find it unobjectionable to postulate brute magnitude relations between physical things and numbers, and who see no great importance in the enterprise of finding thoroughly intrinsic explanations of physical phenomena. For such philosophers, I think that the *only* difficulties with the relationalist position are those posed by field theories. It seems to me however that relationalism is a much more attractive position if it comes with a rejection of heavy duty platonism. In the rest of this essay I will explore some possibilities, consistent with the rejection of heavy duty platonism, for giving a relationalist solution for the problem of quantities in general and the acceleration problem in particular.

9 Geometric Possibility and the Problem of Actuality

A natural reaction when one first hears the problem of quantities is to try to solve it by using the notion of possibility. For instance, a natural

---

[46] Even a vector-valued acceleration functor fails to be independent of the time scale.

proposal for how to define the relation C_2 from congruence and betweenness without going beyond the relationalist's ontology is to say something like

> (P₁) $\exists u(u$ is a geometrically possible point-particle, and u is between x and y, and $xuCuy$ and $uyCzw$),

or

> (P₂) $\Diamond_G \exists u(u$ is a point-particle, and u is between x and y, and $xuCuy$ and $uyCzw$),

or some such thing; where in P_2, the subscript G indicates that the possibility is geometric possibility. I take it that the reason for saying 'geometrically possible' rather than 'logically possible' in P_1 and P_2 is clear: a substantivalist physical theory postulates a specific geometry of space-time; if we are to have a modal surrogate of substantivalism, it must impose the effects of this particular geometry as opposed to other incompatible geometries. Logical possibility can not do this; geometric possibility *can* do it, provided that one lays down suitable empirical axioms that will specify what is geometrically possible and what isn't.[47]

I mention P_1 only to quickly dismiss it: it does not seem to me to be an alternative to the substantivalist's definition of C_2, but merely a transcription of the substantivalist definition into a quite misleading guise. The reason is that in quantifying over 'geometrically possible point-particles' outside the scope of a modal operator, it asserts that such 'geometrically possible point-particles' really do exist (and really do have spatial relations to other existents such as x and y). If they really do exist, the term 'geometrically possible point-particle' is highly misleading: the term 'space-time point' would be better, since it does not try to lull the hearer into failing to realize that they are real existents.

---

[47] I will not discuss here the question of what non-logical axioms on geometric possibility a relationalist could lay down to ensure that it behaved in accordance with the space-time geometry of, say, Newtonian physics. I believe that providing such axioms is not ultimately much of a problem, at least for Newtonian physics. (It may be more of a problem for theories like general relativity, in which the geometry of space-time depends on the distribution of masses, which is not itself fixed by the theory, but I shall not pursue this.) Also, I think that as far as developing a particular theory such as Newtonian physics goes, there isn't really any need to rely on a primitive notion of geometric possibility: for the purposes of the theory, it could be replaced by a defined operator in the definition of which no modal notion beyond logical possibility appears. (Not surprisingly, both the axiomatization of the Newtonian assumptions about geometric possibility, and the definition of a surrogate for geometric possibility suitable for use in developing Newtonian physics, are heavily parasitic on the substantivalist's theory. I'm not sure that that is in itself a criticism of the use of geometric possibility in aid of relationalism, though.)

Clearly, then, P_1 is just substantivalism in disguise. To avoid the problem that besets P_1, we must put the existential quantifier inside the scope of a modal operator. That is what P_2 does.[48]

Nevertheless, there is a serious problem with P_2 as it stands, and that is that in trying to 'modalize away' the commitment to the existence of an entity midway between the point particles x and y, it has 'modalized away' too much: it has 'modalized away' the actual spatial relations between x, y, z and w as well. That is, even if in fact the distance between x and y is *less* than that between z and w, it is *possible* that there is a u between x and y such that xu, uy and zw are all congruent; for it is *possible* that the distance between x and y is twice the distance between z and w, even if it isn't *actually* so. I call this *the problem of actuality*.

How are we to modify P_2 so as to avoid this problem? It does not take much reflection to realize that the problem cannot be avoided using a possibility operator alone. What seems promising, however, is the idea of bringing to bear not only a possibility operator but also what is called an *actuality operator* – an operator that allows you to 'refer back to the actual world' within the scope of a modal operator.[49] For it seems promising that we might solve the problem by saying something like:

> (P_3) $\Diamond_G \exists u (u$ is a point-particle, and u is between x and y, and $xuCuy$ and $uyCzw$, *and the spatial relations between* x, y, z *and* w *are the same as they actually are*).

However, it seems to me very doubtful that anything of this sort will help the relationalist. For, I will argue, how one reads the italicized actuality clause in P_3 depends on whether one is a substantivalist or a relationalist; and only on the substantivalist reading of this clause will P_3 be extensionally correct as a definition of the relation 'has twice the distance of'.

[48] Incidentally, I shall regard modal operators as primitive logical devices, that is, I shall *not* follow David Lewis (1968, 1973) in regarding a sentence of the form '$\Diamond S$' as short for a sentence without a possibility operator that asserts the existence of a 'possible world' in which the sentence S is true. So P_2 not only does not assert the existence of a space-time point renamed (as P_1 did), but it doesn't assert the existence of other dubious entities either.

[49] For a discussion of such an operator, see Crossley and Humberstone (1977). I shall depart from that paper in using 'valid', 'logically true' etc. to mean what they call 'real world valid', 'real world logically true', etc. That is, I shall call 'A iff actually A' logically true (since it can never be false in the real world, whatever the real world is like), even though substitution of 'actually A' for A in *modal* contexts does not always preserve truth.

As a preliminary to arguing this, it is necessary to comment on the term 'spatial relation'. From a substantival viewpoint, the following would naturally be viewed as a spatial relation between x, y, z and w:

R_1: $\exists v(v$ is between x and y, and $xvCvy$ and $vyCzw)$.

It is this relation that the substantivalist uses to define what it is for x and y to be twice as far apart as z and w. On the other hand, from a relationalist viewpoint we would *not* normally think of R_1 as expressing a *spatial* relation between x, y, z and w; this is due to the fact that in the context of relationalism, the v that R_1 asserts to exist would have to be a point of matter rather than a space-time point. This feature of the ordinary use of the term 'spatial relation' is of no great interest in its own right, but it does force us to raise the question of whether we should include R_1 as one of the spatial relations referred to in the italicized clause of P_3. Let us call a reading of 'spatial relation' *broad* if it includes R_1, and *narrow* if it does not.

It is easy to see that a relationalist has to adopt the narrow reading of the term 'spatial relation' in P_3: otherwise, P_3 won't have even an initial chance of being extensionally adequate as a definition of the 'twice the distance' relation, given relationalist assumptions. For if R_1 were to count as a spatial relation, then the only way it could be simultaneously *possible* that

(i) u is a point-particle, and u is between x and y, and $xuCuy$ and $uyCzw$

and

(ii)   the spatial relations between x, y, z and w are the same as they actually are

would be if there were *actually* something v between x and y such that $xvCvy$ and $vyCzw$. (For (i) guarantees that $R_1(x,y,z,w)$ holds within the possibility clause; and (ii) would then guarantee that 'actually $R_1(x,y,z,w)$' holds within the possibility clause, and hence that $R_1(x,y,z,w)$ holds even outside the possibility clause.) But the whole reason for the relationalist having introduced modality was that he or she wanted to get a definition of the 'twice the distance' relation that allowed x and y to have twice the distance of z and w even if there is not anything (point of matter or space-time point) midway between x and y. From a relationalist standpoint, in which all there is is matter, the effect of construing the term 'spatial relations' in P_3 broadly would be to wipe out the modality; and since the non-modal definition R_1 was extensionally inadequate given relationalist presuppositions, then P_3

with 'spatial relations' broadly interpreted would also be extensionally inadequate.

Most of the rest of this section will be devoted to showing that from a relationalist standpoint, P_3 is *also* inadequate if the term 'spatial relation' is read *narrowly*. Before turning to this, however, it is worth asking about the status of P_3 for the substantivalist. If one understands the term 'spatial relations' narrowly even in the context of substantivalism, then the argument that follows against the adequacy of P_3 for the relationalist applies equally against the adequacy of P_3 for the substantivalist. (Of course, this would make no difference to the substantivalist, since he or she unlike the relationalist has the non-modal definition R_1 available.) On the other hand, if one adopts the reading of 'spatial relations' that is more natural in the context of substantivalism – the broad reading – then the argument of the previous paragraph shows that this makes P_3 equivalent to R_1. But from the substantival viewpoint, with its larger ontology, R_1 is a *correct* definition of the 'twice the distance' relation; so P_3, with its modal reference to point-particles, is also correct.

We see then that it *is* possible to define the 'twice the distance' relation via P_3, *if you presuppose substantivalism*.[50] Similarly, you could define the 'twice the distance' relation via P_3 if you assumed a heavy duty platonist version of 'relationalism', for again the modal definition P_3 would be equivalent to a non-modal definition that is adequate given heavy duty platonism. What I will be arguing is that these are the *only* ways that you can define the 'twice the distance' relation via P_3: *the modal definition is only adequate if there is a non-modal definition that is also adequate*. In particular, P_3 is inadequate for the relationalist (who is not also a heavy duty platonist – I will not continue to repeat this qualification). We have already seen that this is so if the term 'spatial relations' in P_3 is construed broadly; so in the various construals of P_3 that follow, I shall construe the phrase narrowly.

In what follows, it will be important to bear in mind that the context in which the relationalist is offering P_3 is in the solution to the problem of quantities. That is, the relationalist is trying to show that certain

[50] This requires a slight qualification. On the substantivalist reading, P_3 is logically equivalent both to R_1 and to 'actually R_1' (i.e., the result of prefixing R_1 with an actually operator). But it behaves like 'actually R_1' rather than like R_1 inside modal contexts (the only contexts where the distinction between R_1 and 'actually R_1' can matter). Consequently, the claim in the text should really be that if you presuppose substantivalism then P_3 gives a definition of the 'twice the distance' relation *that is adequate in all non-modal contexts*. (See fn. 49. See also fn. 57, where the sort of point made here is made in a different way, in connection with a certain version of modal relationalism.)

relations which are unproblematic for the substantivalist are also available to the relationalist. In particular, the relationalist whose primitive geometric relations are just congruence and betweenness is trying to show that the 'twice the distance' relation is available to him or her, and is offering P_3 to show that this relation is indeed available. With this context in mind, let's turn to different interpretations of the italicized clause in P_3, to see if any of them will work given this context.

Let's start with the weakest reading of P_3. We are assuming that the modal relationalist's primitives are betweenness and congruence. (If not, we may have to pick another example than the 'twice the distance' relation to illustrate the problem of quantities and the proposed modal solution to it.) So a natural way to understand the italicized phrase '*the spatial relations between x, y, z and w are the same as they actually are*' is as the conjunction of all claims of the form

a is between b and c iff actually (a is between b and c)

or

a,b congruent to c,d iff actually (a,b congruent to c,d)

in which one of x, y, z and w is substituted for both occurrences of a, and the same or another of x, y, z and w is substituted for both occurrences of b, etc. But clearly if we explicate the italicized clause in this way, P_3 is much too weak. Suppose for instance that in actual fact, x, y, z and w are situated on a straight line as follows:

situation A

$$x{-}-----{-}y{-}------------z{-}--{-}w$$

3 ft.   4 ft.   1 ft.

Then the actuality clause as construed above restricts attention to possible situations in which x, y, z and w are on a straight line in that order; but it imposes virtually no restrictions on their relative positions on that line; in particular, it is geometrically possible consistently with the actuality clause so construed to have:

situation B

$$x{-}----y{-}------------z{-}--{-}w$$

2 ft.   4 ft.   1 ft.

Similarly it is geometrically possible consistently with the actuality clause to have a situation B* just like situation B except with an object u midway between x and y. So P_3 will come out true in virtue of the possibility of situation B*, even if situation A actually holds; therefore P_3 as we are construing it is inadequate as a definition of the 'twice the distance' relation.

Let's try a second construal of the actuality clause. The trouble with the first construal, it might be suggested, is that it only allows us to hold fixed those relations in the actual world that the relationalist has taken as primitive; whereas we ought to allow that all the narrow relations definable from these are held fixed. In other words, we should take the italicized clause in P_3 as the infinite conjunction of all the formulas of form

$$\phi(x,y,z,w) \text{ iff actually } \phi(x,y,z,w)$$

in which no reference is made to any objects between x and y.[51] (We can imagine that some device of infinite conjunction, such as a 'substitutional quantifier', is added to the relationalist's language so that the infinite conjunction can be expressed in a finite notation.)

It is clear however that this second construal is really no more powerful than the first: if we can't distinguish situation B from situation A in terms of the primitives (betweenness and congruence, we're assuming), then obviously we can't distinguish them in terms of any relations first order definable out of the primitives; therefore we can't distinguish them in terms of any totality of such relations, even if the totality be infinite. Any plausibility that the suggestion may have had comes from thinking that we ought to be able to define in the theory some formula $\phi(x,y,z,w)$ which restricts x and y to being twice the distance of z and w. Well, we *can* define such a formula if we assume a substantivalist ontology (and allow reference to space-time points between x and y); but if we could define such a formula with only a relationalist ontology then we would have solved the problem of quantities without modality. The invocation of modality – even with actuality, at least as the actuality clause of P_3 is understood on our first two construals of it – gives definitions that are either extensionally incorrect or unnecessary. For a substantivalist they are unnecessary: they may be extensionally correct, but they don't help in solving the problem of quantities since that problem was solvable without them. (Similarly for someone who maintains 'relationalism' by adhering to heavy duty platonism.) But for a relationalist who rejects heavy duty

---

[51] This last restriction is the simplest way to rule out 'broad spatial relations' like R_1 – as, you will recall, we must do in order to prevent their being a 'spatial relation' of x, y, z and w which holds in situation B* but not in situation B, and thus to prevent the actuality clause failing in situation B* even when situation B is actual. The argument to follow will not depend on the details of how the restriction is formulated. (Presumably, any restriction designed to accord with the intuitive idea of a *spatial* relation on a relationalist view will rule out at least as much as I have ruled out formally, and that will be enough to make the objection that follows apply.)

platonism, P_3 is simply incorrect, at least on our first two construals of the italicized phrase in it.

Let us now try a third construal. On the second construal, the actuality clause was in effect taken as

(*) $\Pi R[R(x,y,z,w)$ iff actually $R(x,y,z,w)]$,

where Π is a substitutional quantifier understood in such a way that the substituends can not themselves include substitutional quantifiers. (For simplicity of formulation I leave tacit the required restriction to 'narrow' relations discussed in the previous footnote.) We could of course introduce a hierarchy of substitutional quantifiers, each allowing the ones of 'lower level' to appear in its substituends; but doing so, and allowing the substitutional quantifiers in (*) to be of high level, would not improve the prospects of (*). Alternatively, we could mimic Kripke's modification of the Tarski hierarchy of truth predicates (Kripke 1975), and stick to a single substitutional quantifier but let it be a quasi-impredicative one in which formulas that contain the substitutional quantifier are allowable substituends. Doing this requires that we allow substitutional quantifications to have truth value gaps, just as truth value gaps are required in Kripke's work on truth. (The possibility of mimicking Kripke's construction for truth in the case of substitutional quantification was noted by Kripke himself in footnote 31 of his paper.)

So a third construal of the actuality clause would be again as (*), but this time construing the substitutional quantifier in the more liberal, quasi-impredicative, fashion. But this won't work either: however we specify the details of the construction (e.g., whether we choose a Kleene three-valued logic or choose supervaluations to handle the truth value gaps), either (*) will turn out true of situation B* when (as before) situation A is assumed as actual, or (*) will turn out truth valueless in situation B* even when situation B is assumed as actual. (The second failing is the one that arises on most ways of specifying the details; it results in P_3 giving a definition of the 'twice the distance' relation that isn't true when it should be, rather than one that is true too often.) That we should always have one or the other of these failings is not surprising, since situations A and B are isomorphic with respect to the relationalist's language.

The obvious suggestion as to how to avoid the problem common to the first three construals of the actuality clause is to replace the substitutional quantifier in (*) by a genuine quantifier over physical relations – where 'relations' does not mean relations in extension, i.e., sets of ordered n-tuples, it means instead multi-place properties. It is worth remarking that to make this move is to take a substantial step: hitherto in this essay (except when expounding higher order monadicism or the related doctrine in footnote 44), any talk of properties or relations that I have indulged in was a mere manner of speaking; but the proposal now under consideration is that to avoid postulating space-time, we should instead genuinely postulate physical properties and relations (and accept a primitive notion of geometric possibility and an actuality operator to boot). It seems to me surprising that anyone would regard properties and relations as less problematic entities than space-time regions; but I will not pursue this matter here.[52] Instead, I will confine myself to whether the postulation of physical properties (including multi-place physical properties, i.e., physical relations), and the replacement of (*) by

$$(**) \quad \forall R[\Delta(R,x,y,z,w) \text{ iff actually } \Delta(R,x,y,z,w)],$$

will solve the problem of actuality. ('$\Delta(R,x,y,z,w)$' means that the relation R is instantiated by x, y, z and w in that order.) Again, the property quantifier is tacitly restricted to narrow spatial relations, taken to mean at the very least relations that don't involve questions about the existence of things between x and y. The reason, again, is that we need (**) to come out true in situation B* if situation B is actual.

The details of the answer depend on just how one conceives of properties and relations. I believe that the most widespread conception of physical properties and relations is a predicative one. On this conception (advocated, for instance, by Hilary Putnam (1970)) we have a few *physically basic* properties and relations; from these, all other properties and relations are constructed. The construction proceeds by levels. At the first level, no reference to any totality of physical properties is allowed in the construction of a new property; at the second level, reference to the totality of first level properties is allowed; at the third level, reference to the totality of first and second level properties is allowed; and so on. Every physical property appears at some level of the hierarchy, according to this picture. This is a popular picture; for instance, it is presupposed by much of the functionalist literature in the

---

[52] For some remarks comparing the status of space-time regions with that of *mathematical* entities, see essay 2, pp. 67–73. For whatever it's worth, it seems to me that physical properties are a little less problematic from an epistemological standpoint than mathematical entities, at least when, as here, one is applying no higher order predicates (beyond an instantiation predicate) to them, but that they are a good bit more problematic than space-time regions.

I would concede that properties as used here are a bit less problematic than properties as used by a higher order monadicist or by the advocate of the theory HOM discussed in fn. 44, since in those latter cases the use of higher order predicates (beyond instantiation) made it especially hard to see why there should be a principled distinction between 'properties' and 'individuals'.

philosophy of mind, where functional properties are (implicitly if not explicitly) viewed as properties of levels greater than or equal to two. For to say that pain is a functional property means that to be in pain is to have *some physical property or other* which has such and such functional role. The property which has the specified functional role (the property *in virtue of which* the given individual is in pain) is allowed to be different from one pain-feeler to another. It seems somehow 'circular' to say, when asked what the physical property in virtue of which a given pain-feeler is in pain, that it is simply the property of being in pain; rather, we must specify a *non-functional* physical property, or anyway a *less* functional one, that the individual has in virtue of which it is in pain. The idea that this must be done to avoid circularity reveals a conception of physical properties as falling into levels in the way just described: one property is 'less functional' than another if it is of lower level, and a property is non-functional if it is of level one.[53]

On this relatively attractive conception of properties, how does P_3 fare if the actuality clause in it is explicated as (**)? The answer is that it has an almost identical failing to the one that it had on the second construal of the actuality clause. It would in fact be a completely identical failing if one assumed that the basic physical properties from which all others were constructed are betweenness and congruence, and that the means by which new properties are constructed at each level from already available properties is first order definability. In that case, there would be no real difference between the present proposal and the one involving purely predicative substitutional quantifiers (of level ω). So even after the shift from (*) to (**), P_3 would come out true of situation B* even if the actual situation is A, and consequently P_3 would fail to define the 'twice the distance' relation.

How about if we remember that the basic physical properties might not be betweenness and congruence and/or that the allowable means of construction might be broader than first order definability? (E.g., we might allow cardinality quantifiers or other apparatus that is consistent

---

[53] Actually, functionalism presupposes a bit less than this: it doesn't actually require that we be able to attach levels *to properties* in this way, but only that we be able to attach a level *to each instantiation of a property*. Different instantiations of the same property might get different levels. The 'quasi-impredicative' conception of properties to be mentioned shortly weakens the predicativity requirement in just this way.

Also, I don't mean my remarks to suggest that functionalism requires genuine quantification over properties (as opposed, say, to substitutional 'quantification'); my point is only that *if* functionalism is viewed in terms of genuine quantification over properties, *then* we should view the properties quantified over in a predicative or quasi-predicative way.

with predicativity requirements.)[54] This really wouldn't change anything of principle. Recall that the problem of quantities was a problem that apparently arises for a non-modal relationalist *whatever* the choice of primitive predicates for his or her theory. (The problem might not arise in the context of defining the 'twice the distance' predicate, of course, for the 'twice the distance' predicate could simply be taken to be one of the primitive predicates – but it would arise in the context of defining *some* predicates, and the problem of defining the 'twice the distance' predicate was simply used to illustrate the kind of failure that we would get in general.) I don't know how one would prove conclusively that whatever the choice of primitive predicates, the problem of quantities will arise in some form, but I think that the discussion in section 7 should make that pretty clear; in any case, if there *is* a choice of primitives for which the problem doesn't arise, that would show that a reasonable *non-modal* solution to the problem of quantities is available, and hence would be of no relevance to our present discussion of whether there is a way to use modality to solve the problem. But if the problem of quantities arises for the non-modal relationalist for any choice of primitive predicates, then it would arise even if the predicates in the language stood for 'the physically basic properties and relations'; consequently, the discussion of the previous paragraph would still apply.

A similar point holds if we allow that the construction of properties at each level proceeds not by first order definability but by definability of a broader sort, say definability in a language containing quantifiers like 'there are infinitely many.' Given any such proposal, we have to ask first whether the problem of quantities would still arise for a *non-modal* relationalist who allowed the new devices of definition into the language in which he or she developed his or her theory; conceivably, the proposal to add the new devices of definition will suffice for solving non-modally the problem about definability and representation theorems, and we won't need to introduce modality at all. (Of course, the 'infinitely many' quantifier doesn't have a remote prospect of doing this in any of the examples of the problem of quantities that we've seen; but it is hard to prove that *no conceivable addition* to the logical devices available will solve the problem.) In any case, what is presently at issue is whether the introduction of *modality* could help solve the problem of quantities; since that is a problem which won't even arise if the problem is already solvable non-modally, we can assume for present purposes that no satisfactory non-modal solution would be possible even if we were to

---

[54] One thing we *couldn't* allow is definability via impredicative second order quantifiers, if the second order quantifiers were viewed as ranging over physical properties rather than (as usual) over classes.

develop science in an expanded language that includes logical devices corresponding to the ones by which 'properties are really generated'. The previous arguments against P_3 (or against analogous definitions of concepts other than C_2) based on the first three construals of its actuality clause still go through, and since there is now no longer a disparity between the assumed methods of property generation and the contemplated language of science, the problems that arose for the second construal will apply to the fourth as well.

Are there impredicative views of properties on which (**) would fare better as an interpretation of the actuality clause in P_3? First let's consider briefly how quasi-impredicative views of properties would fare. By a quasi-impredicative view I mean one that uses an idea of partially defined properties, conceived of in accordance with a construction similar to Kripke's construction for truth, to achieve the spirit of predicativity requirements without attaching levels to properties but instead attaching levels to instantiations of properties (with different instantiations of the same property sometimes getting different levels). It is not difficult to come up with such a construction,[55] and the resulting view has the advantage of still according with the functionalist paradigm and yet having some added flexibility: e.g., in allowing instantiations of functional properties to 'seek their own level' (to adapt a phrase from Kripke's work on truth) rather than arbitrarily cutting off those above a given level; and in allowing the existence of a property of being a property. But it is not necessary to pursue it further here; it is clear that just as the difficulties with using a fully predicative version of substitutional quantification in an interpretation of the actuality clause carry over to difficulties in using a fully predicative version of properties, so also the difficulties with using the 'quasi-impredicative' version of substitutional quantification in the actuality clause (as in construal three) carry over to the use of quasi-impredicative properties. P_3 won't work on this construal either: as on all of the previous ones, it simply gives the wrong extension to the predicate 'twice the distance' that it was supposedly defining.

Let us turn finally to a fully impredicative conception of properties, one in which the whole idea of their being defined from previously available properties is abandoned. (Properties on this view behave rather

---

[55] A similar construction for sets, conceived of as supplying an alternative to the iterative conception of sets, was developed by Gilmore (1974); and a similar construction for proper classes, conceived of as things to be added on to an iterative conception of sets, was given in Maddy (1983). A generalization to 1-place physical properties is obtained by giving up extensionality and restricting comprehension at the lowest level to physically basic vocabulary, and the generalization from that to multi-place properties is also pretty routine.

like sets on the iterative conception of sets; they presumably fall into types (otherwise a property-theoretic analogue of Russell's paradox is almost inevitable), but the types are based only on what entities they apply to, not on any definition they might have.) This conception, while natural for sets, seems to me to have little to recommend it in connection with physical properties, as my earlier remarks on functionalism probably make clear.[56] But let us see whether if we adopted it, and then used (**) to explicate the italicized clause in P_3, the resulting construal of P_3 would suffice for defining the 'twice the distance' relation from the relationalist standpoint.

The answer is still no, but this time the reason is interestingly different from on the previous construals of the actuality clause. In the case of the previous construals, the problem was that the proposed relationalist definition of the 'twice the distance' relation ended up with the wrong extension. (Or to be more accurate, the proposed relationalist definition of *some* needed predicate ended up with the wrong extension; the 'twice the distance' relation was an illustration, chosen on the assumption that the geometric predicates which the relationalist took as basic were just betweenness and congruence.) If the relationalist explicates the italicized clause of P_3 by (**), and adopts a thoroughly impredicative conception of properties, then we can no longer assert that P_3 would give to the 'twice the distance' relation the wrong extension. But the relationalist is still in trouble. For (to put it roughly) the relationalist needs to assert, but can't, that he or she would be giving to the 'twice the distance' relation the *right* extension.

We have already seen that if one explicates the italicized clause in P_3 at face value, along the lines of (**), then the question of the extensional adequacy of P_3 turns on the range of the relation quantifier in (**): if this quantifier ranges only over narrow relations predicatively constructible from betweenness and congruence, P_3 is extensionally too broad, whereas if it includes broad relations (whether predicative or impredicative) this difficulty is avoided but there is the opposite difficulty that if one assumes the relationalist's ontology then P_3 is extensionally too narrow. Perhaps if we construe the quantifier impredicatively but exclude broad relations, the extension will turn out just right; but if so, this will have to be *shown*. But there is at least a *prima facie* difficulty in seeing how it *can* be shown (even allowing the use of impredicative reasoning in the showing). The only obvious way to guarantee that there is a (narrow) spatial relation that x, y, z and w stand in in situation

---

[56] Note also that one possible advantage of the quasi-impredicative conception – the existence of a property of being a property – is lost on the fully impredicative view, due to the presence of types in the latter.

B* but not in situation A (or equivalently, that there is some spatial relation that they stand in in situation B but not in situation A) is to assume that the 'twice the distance' relation is itself such a spatial relation. But the relationalist is in no position to assume that in this context, for since the 'twice the distance' relation is the relation he or she claims to be defining, the needed assumption is precisely the claim he or she wants to prove.

Put this way, the objection sounds a bit like a general objection to impredicative definition that is invalid. Let's look at a classic example of an impredicate definition, the definition of what it is for an ordinal number to be finite:

$$\text{Fin}(\propto) \longleftrightarrow \forall P[P \text{ is inductive } \& \ P(0) \supset P(\propto)]$$

where 'P is inductive' is defined to mean

$$\forall \beta[P(\beta) \supset P(\beta+1)].$$

The invalid objection is that on this definition one could never know that a given number, say 2, was finite, since to show that it was you would need to show that 2 had *every* inductive property of 0, and this requires showing that if finitude is an inductive property (as in fact it is) then 2 is finite. So, the objection goes, you couldn't show that 2 was finite without first showing that 2 was finite; and the same will hold for any other impredicative definition. The reply, of course, is that there is no difficulty in showing that *if finitude is an inductive property then* 2 is finite: for 0 is clearly finite by definition, and that implies that if finitude is inductive then 1 is finite, and that in turn implies that if finitude is inductive then 2 is finite. What holds for finitude holds for any other property, and this establishes that 2 is finite. We don't need that 2 is finite in establishing this, but only that *if finitude is inductive, then* 2 is finite; so contrary to the perhaps initially plausible objection, it is possible to establish facts about the extension of impredicatively defined properties.

My reason for reviewing this elementary point is to provide an example of the sort of proof one *can* sometimes give for impredicatively defined concepts, which avoids an apparent circularity, and my point is that such a proof would be *needed* (and, I will argue, is *unavailable*) in the case of an impredicative definition of the 'twice the distance' relation. To review the situation: if impredicative definition is allowed, then P_3 (understood impredicatively) defines a certain relation; putting things roughly and intuitively for the moment, what we need to know is whether this relation is the 'twice the distance' relation. In order to establish that this is the 'twice the distance' relation, one would need that it doesn't hold of x, y, z and w in situation A. That requires showing that there is some narrow relation R such that

(#) not[$\Delta(R,x,y,z,w)$ iff actually $\Delta(R,x,y,z,w)$)]

holds of situation B* (or, equivalently, of situation B) on the supposition that the actual situation is A. Now, it is perfectly legitimate, when one has accepted impredicative definitions, to allow the relation defined by P_3 to be the needed relation R. The only requirement is that one be able to *prove* that with such a choice of R, (#) holds. But, I claim, we can not prove this instance of (#).

The easiest way to see this is to note that if we assume that the possibility operator obeys the laws of S5 then in fact we can prove the opposite! For a fundamental feature of an actuality operator in S5 is that 'it is actually the case that it is possible that A' is *strongly* equivalent to 'it is possible that A', in the sense that if one be substituted for the other, even in modal contexts, the result of the substitution is equivalent to what one started with. Consequently, since P_3 begins with a possibility operator,

$\Box_G[\Delta(R,x,y,z,w)$ iff actually $\Delta(R,x,y,z,w)$)]

holds when R is the relation defined by P_3; so (#) can't possibly hold of situation B for the relevant instance of R, whatever situation is actual.

The argument for the *disprovability* of the relevant instance of (#) rests on the assumption that the possibility operator in question obeys all the laws of S5. That assumption seems to me to be undeniably correct; but rather than argue this, let me simply note that any weakening of the modal logic wouldn't really be of much help. To be sure, such a weakening (say, to S4) would *remove the disproof* of the relevant instance of (#). But it couldn't serve to make the relevant instance of (#) *provable*, and that is what the relationalist needs. So unless one can argue that the possibility operator in question obeys laws *incompatible with* the S5 laws – surely not a plausible claim – then there is no hope whatever of the relationalist finding a way to argue for the relevant instance of (#). (And even if one were to use a modal logic incompatible with S5, it is completely unobvious how an argument for the relevant instance of (#) could possibly go.)

As I've said, I think that the S5 laws are clearly the relevant laws of modal logic in the present context, and if this is right then the relevant instance of (#) is not only unprovable, it is disprovable.[57] This still doesn't show the extensional inadequacy of P_3, for we haven't ruled out the possibility that there is some narrow property R *other than* P_3 (perhaps a property not definable even impredicatively in the relationalist's language) for which (#) holds; but the only initially plausible impredicative strategy for showing that there *is* such a relation has been shown unworkable.

It might be objected that it is just *obvious* that there is a relation R for which (#) holds: namely, the 'twice the distance' relation. Well, of course it is obvious (putting aside any general doubts about quantifying over relations). The question is, in granting the obviousness of the existence of this and the infinity of similar relations that would be needed for defining other important parts of our space-time ideology, do we presuppose the truth of either substantivalism or heavy duty platonism. The above argument shows in effect that we do presuppose this.

This last point can be made most clearly if I respond first to another objection. My presentation of the argument in this section has been rather informal, in the following sense: I have spoken as if certain situations such as situation A were given, and asked how the modal relationalist establishes that his or her definitions do or do not apply to those situations. It could rightly be objected that this is not a good way to look at things: it is not clear what it would mean for a situation to be 'given'. A theorist *describes* various situations in his or her own language; but if a theorist simply defines the 'twice the distance' relation by P_3, then trivially there is no possibility of being able to describe a

---

[57] Indeed this argument can be turned into a proof that even if P_3 is extensionally adequate in the sense that whatever the actual world is like, the relation holds of the actual world when it should, still the relation won't have the right modal properties. For the sentence

$$\Diamond_G \exists x, y, z, w \text{ (the distance between } x \text{ and } y \text{ is twice that between } z \text{ and } w,$$
even though actually the distance between x and y is more than twice that between z and $w)$

ought to come out true, but it won't if the 'twice the distance of' relation is defined as above (or if it is defined by any formula whose main connective is a modal operator). (If this is puzzling, note that extensional adequacy even in the general sense just defined does not guarantee that whatever the actual situation, the defined term comes out true of non-actual situations just when it should; see fn. 49, and also fn. 50.)

I won't press this point, for it could be maintained that sentences like the one displayed in the previous paragraph aren't interesting enough to worry about. This has a certain plausibility, though it is rather awkward to maintain it at the same time that one is exploiting modal operators and actuality operators to the extent that the modal relationalist is doing.

situation where P_3 and the 'twice the distance' relation as he or she defines it diverge.

I hope it is clear that this objection does not cut against anything other than the rather loose and intuitive way that I have chosen to formulate the difficulty with P_3. A more formally correct way of putting the difficulty with P_3 is as a difficulty with *proving the properties of the 'twice the distance' relation that need proving*, if that relation is defined via P_3. For instance, suppose we define the 'twice the distance' relation by P_3, and define the 'three times the distance' relation analogously. Then one thing that we had better be able to prove is that if the distance between x and y is three times that between z and w, then it is not also twice the distance between z and w (unless z and w occupy the same point). And the difficulty that I have described intuitively in terms of situations A and B and B* is the difficulty of seeing how that is to be proved if one does not have substantivalism or heavy duty platonism at one's disposal. [To say that xy has both twice the distance of zw and three times the distance of zw would be to say that it is possible that there is a u between x and y such that xu, uy and zw are all congruent and the actuality clause holds, and also that it is possible that there is an s between x and y and a t between s and y such that xs, st, ty and zw are all congruent and the actuality clause holds. If we could somehow 'combine these possibilities into one' – i.e., argue that it is possible that there are u, s and t satisfying all of the above conditions including the actuality clause – then we could argue that in the possible situation in question x and y have to occupy the same point, and hence so do z and w; and hence (by the actuality clause) that z and w must actually occupy the same point. But there is no way to 'combine the possibilities into one': even if we could find a way to get a possibility where u, s and t all exist together, we couldn't argue that in this possible situation they have the properties they have in the two other ones. (The actuality clause won't help here: recall that the quantifier is restricted to narrow relations of x, y, z and w: properties of the intermediate objects u, s and t needn't be preserved.)]

Of course, one way to prove the needed law is to just add it as an axiom. ('After all', one might say, mimicking the objection of three paragraphs back, 'it is just obvious.') There could be no objection to doing this if this were the only axiom one needed to add (just as there could be no objection to adding the 'twice the distance' relation as a primitive if that were the only new primitive one needed to add). The problem, however, is that this axiom is far from enough, since the difficulty presented by situation A is just one of a large class of difficulties. To begin with, for each pair of distinct predicates 'the distance between x and y is r times that between z and w' and 'the distance between x and y is s times that between z and w', one will need an axiom saying that if the first holds the second fails (except when z and w occupy the same point). In fact, one will need a great deal more than this even to get the right laws about how these distance comparison predicates interrelate (as is easily seen by the fact that the above axioms do nothing towards inducing the natural ordering among these distance comparison predicates).[58] Moreover, as I stressed in sections 6 and 7, the distance comparison predicates are only a tiny part of the problem of quantities; if the problem of quantities is to be treated modally, then modal definitions analogous to P_3 would have to be introduced in every case, and virtually all of the required interconnections between these different concepts would have to be given as separate axioms. It is not much of an exaggeration to say that everything that the substantivalist or heavy duty platonist derived as a theorem, the modal relationalist would have to take as a separate axiom. The result would not be a theory at all in any interesting sense.

I should emphasize again that the difficulty just reviewed is not really a difficulty about the modal characterizations provided by P_3 and its ilk: that is, *in certain circumstances* the required interrelations among these characterizations do *not* all need to be taken as separate axioms, they can be proved. But the circumstances in which this can be done are precisely that there *also* be *non-modal* characterizations of the same concepts (whether provided by a substantivalist theory or a heavy duty platonist 'relationalist' theory). The modal characterizations are adequate precisely when they are not needed to solve the problem of quantities. For that reason, they are of no help in solving that problem.

10 Counterfactuals and Possible Worlds

I think that the previous section makes it pretty clear that one cannot solve the problem of actuality by invoking an actuality operator alone (even in conjunction with a quantifier over properties, however properties are construed). This does not totally rule out all hope of a modal solution to the problem of quantities: for it is conceivable that one might find further operators besides an actuality operator which would enable us to solve the problem that the actuality operator alone can not solve. What such devices might be like I am not sure, but I do not know how to prove that none can be found.

---

[58] Another of many examples of laws that would have to be added is the instances of the triangle inequality: for each $k>i+j$, the sentence 'If xvC_izw and vyC_jzw and not $(zwCzz)$ then not (xyC_kzw).'

One suggestion that might appear natural is to use counterfactuals. I can not fully consider this idea here, but would like to note three things.

First, counterfactuals are, notoriously, extremely vague. This does not hamper their use in certain contexts; but it does seem to me that it makes it unattractive to rely on them in the formulation of physical theory. If so, then since we need geometric notions in the development of physical theory, we shouldn't use counterfactuals in our account of geometric notions.

Second, it is a rather widely held principle that counterfactuals can never be (in Michael Dummett's words) 'barely true': that is, if a counterfactual is to be true, there must be some facts (known or unknown) statable without counterfactuals in virtue of which the counterfactual is true. (See Dummett 1976, p. 89). Obviously, if this principle is accepted then a relationalist can not use counterfactuals to solve the problem of quantities; for the principle requires that if distance relations are defined counterfactually, then situations that differ in their distance relations (like situations A and B) must differ in some non-counterfactual respects as well. A substantivalist would of course grant that they differ in non-counterfactual respects, but if a relationalist could find a way to grant this he or she would have no need of counterfactuals.

These first two points make the idea of appealing to counterfactuals to solve the problem of quantities seem most unappealing. The third point is that it also seems unworkable. It is unworkable for a reason very similar to the reason why the impredicative property version of P_3 was unworkable: it appears that no *theory* about counterfactually defined relations will be forthcoming, unless those relations can also be defined non-counterfactually. (In effect, this is a formal motivation for the Dummettian metaphysical principle that counterfactuals can't be barely true.) For instance, to prove the incompatibility of the 'twice the distance' relation and the 'three times the distance' relation (given that the objects z and w in the last two places of the relation do not occupy the same point, i.e., given that zw is not congruent to zz), you'd need the incompatibility of

(a) if there were a point u midway between x and y, then uy would be congruent to zw

and

(b) if there were a point s between x and y and a point t between s and y such that xs, st and ty were all congruent, then ty would be congruent to zw.

And if these counterfactuals are somehow derivative on non-counterfactual claims – e.g., claims about space-time points – then presumably by considering the way in which they are derived together with the relations that can be proved to hold among the non-counterfactual claims, one *could* get an argument for the incompatibility of (a) and (b). But if we take the counterfactuals as having no non-counterfactual backing, it is hard to see how the incompatibility of (a) and (b) is to be argued; apparently, it must be taken as a brute assumption. That wouldn't be so bad if it were one of a small class of brute assumptions we needed to make. But as we saw in the penultimate paragraph of section 9, this is just one example of a *huge* class of assumptions that a relationalist would have to take as brute. Again, it would not be much of an exaggeration to say that everything that the substantivalist derived as a theorem, the counterfactual relationalist would have to take as a separate axiom. The result could hardly be called a theory at all.

There is another possible version of modal relationalism I now want to consider briefly, which does not invoke counterfactuals (or even a modal operator!), but instead involves

> (i) enriching one's ontology to include possible worlds construed in the way that David Lewis (1973, section 4.1) construes them – i.e., as universes rather like the actual universe, and containing no spatio-temporal relations to our universe (just as past time-slices of our universe are rather like the current time-slice, and contain no *spatial* relations to the current time-slice),

and

> (ii) invoking a predicate of congruence that applies across worlds.

If both these moves are made, then it is no trick to avoid the problem of actuality. For then we can avoid using a possibility operator entirely: for instance, to define xyC_2zw, we simply say,

> (P_4) there is a possible world P in which the geometry of Newtonian space-time holds and in which there are entities x^*, y^*, z^*, w^* and u^* such that u^* is between x^* and y^*, $x^*u^*Cu^*y^*$, $u^*y^*Cz^*w^*$, x^*y^*Cxy, and z^*w^*Czw.

(The last two clauses, invoking crossworld congruence, correspond to the actuality clause in the modal definition P_3.)

I take it however that this sort of solution to the problem of quantities has very little to recommend it. For the main point of relationalism was to avoid the ontology of space-time regions. It is hard to imagine anyone with ontological scruples about space-time regions thinking that Lewis-type possible worlds are any better. (There may also be problems with invoking 'crossworld congruence', even given a Lewis-like conception of possible worlds, but I will not pursue them; the problems with (i) are enough.)

Of course, there are many contexts in which talk of possible worlds is quite harmless: in those contexts, such talk is simply a heuristically useful way of saying things that could be said with a possibility operator, or a possibility operator and actuality operator together (or perhaps even a possibility operator and an actuality operator and a counterfactual connective). In the present context, that is not so: we are invoking possible worlds precisely because we couldn't solve the problem with a possibility and actuality operator alone. Perhaps we can find some additional operator such that when it is invoked along with possibility and actuality operators, the reference to possible worlds can be translated away and thus shown harmless. Perhaps; if so, then it is the use of that additional operator, not the invocation of possible worlds, that would be the solution to our problem.

My next point is that even among views that take quantification over possible worlds at face value (as opposed to regarding it as translatable away into modal talk), there are important differences. For instance, Robert Stalnaker (1976) has proposed a relatively appealing alternative to Lewis' conception of possible worlds: regard a 'possible world' as merely a property that the universe might have had. So, if we are willing to put aside doubts about quantification over properties, can we accept P_4 on this construal of possible worlds? Unfortunately, it does not seem possible to make sense of anything like P_4 on Stalnaker's conception of possible worlds.

To see this, note first that it doesn't make sense to talk of objects 'existing in' properties; so talk of 'a possible world in which there are objects x^*, y^*, z^*, w^* and u^* such that $F(x^*,y^*,z^*,w^*,u^*)$' must be understood as talk of a property Q such that it is necessary (in some sense of 'necessary' whose status I will not pursue) that if the universe has Q then there are x^*, y^*, z^*, w^* and u^* such that $F(x^*,y^*,z^*,w^*,u^*)$. But given this understanding of possible world talk, how are we to understand the crossworld congruence claims involved in P_4? We can't say

> (P_4^*) there is a property Q such that $\Box_p$ [if the universe has Q then the geometry of Newtonian space-time holds and there are x^*, y^*, z^*, w^* and u^* such that u^* is between x^* and y^*, $x^*u^*Cu^*y^*$ and $u^*y^*Cz^*w^*$], and x^*y^*Cxy and z^*w^*Czw

(where $\Box_p$ is the relevant sort of necessity); for the occurrences of x^*, y^*, z^* and w^* in the last two conjuncts are not bound by quantifiers, as they clearly need to be. But there is no way to modify P_4^* so as to bind them, without either (i) moving the x^*, y^*, z^* and w^* quantifiers to outside the modal operator, or (ii) moving the clause 'x^*y^*Cxy and

z^*w^*Czw' to inside the modal operator. The latter course would clearly be disastrous: it would result in a modification of P_4^* that said only that there was a property Q such that, necessarily, the universe has Q if and only if both (a) xy has twice the distance of zw and (b) there is a point-particle u^* either midway between x and y or midway between some other pair of points x^* and y^* such that x^*y^*Cxy. But such a property exists (uninstantiated of course) in situation A as well as in situation B, so that can't be what we want to say. And we can't add the further claim that the universe has this property Q, for that fails in situation B as well as in situation A.

Moving the x^*, y^*, z^* and w^* quantifiers outside the modal operator is the option that remains. Adopting it, we get

> (P_4^{**}) there is a property Q and point-particles x^*, y^*, z^* and w^* such that
> (a) $\Box_p$ [if the universe has Q then there is a point-particle u^* such that u^* between x^*y^* and $x^*u^*Cu^*y^*$ and $u^*y^*Cz^*w^*$], and
> (b) x^*y^*Cxy and z^*w^*Czw.

But this still won't work: in situation A, we can take x^*, y^*, z^* and w^* to be x, y, z and w respectively, and both clauses will hold. (Clearly, things are not improved if we drop the quantifiers over x^*, y^*, z^* and w^* and replace these variables by x, y, z, and w throughout P_4^{**}.)

One final attempt suggests itself:

> (P_4^{***}) there are properties Q_1 and Q_2 such that
> (a) $\Box_p$ [if the universe has Q_1 then there is a point-particle u between x and y such that $xuCuy$ and $uyCzw$];
> (b) the universe has Q_2; and
> (c) Q_2 is like Q_1 as far as the spatial relations between x, y, z and w are concerned.

But of course, this raises just the problems about the interpretation of the phrase 'spatial relations' discussed at length in section 9. That is, if the term is interpreted broadly, so that R_1 counts as a spatial relation, then the universe can't have Q_2 unless there is actually a particle midway between x and y; so P_4^{***} will fail in situation B as well as in situation A. On the other hand, if 'spatial relations' is interpreted narrowly, then we can not establish that the universe fails to have Q_2 even in situation A. The upshot is that the introduction of Stalnaker-type possible worlds is of no help whatever in providing a modal solution to the problem of quantities.

It seems that almost wherever we turn, the same problems arise. As I've said, I will make no attempt to prove that there is no reasonable way to solve the problem of quantities modally; rather, the aim of these last two sections of the paper has simply been to pose a challenge to the modal relationalist to say how these problems are to be circumvented. I do not believe that the challenge is at all easy to meet, but I will not pursue any further attempts to meet it now.

Postscript

Three small addenda to the argument of the last two sections of this essay.

1   Harold Hodes (1984a) has introduced a variant of the actuality operator, which instead of 'temporarily undoing the effect of' all the modal operators in which it occurs, 'temporarily undoes the effect of' only the last one. I'll call it the backspace operator, since as Hodes says it behaves like the backspace key of a typewriter whereas the standard actuality operator behaves like a carriage return. The backspace operator (which I didn't know about when I wrote the paper) is at least as natural a reading of 'actually' as is the more standard carriage return operator, and is more flexible. Would it have helped in avoiding the problems of the modal substantivalist? Only slightly. Had I used the backspace operator instead of the standard operator, the slightly anomalous behaviour of modal definitions of spatio-temporal relations noted in footnote 57 would not have arisen. But very little weight was put on that in the paper anyway; and the main argument against modal definitions of these relations would be completely unaffected by a shift from the standard operator to the backspace operator.

2   In the discussion of possible worlds in section 10, I parenthetically remarked that even if we put aside doubts about Lewis' conception of possible worlds, there may be problems with invoking 'crossworld congruence'. What I mainly had in mind as a problem with crossworld congruence is that (as my footnote 15 on the Leibniz argument suggests) I sympathize with the idea that if you talk in terms of possible worlds you should do so in a way that makes 'qualitatively isomorphic' worlds identical, and this is naturally construed as ruling out the existence of two worlds which differ only in that the distances in one are twice what they are in the other. This would rule out using crossworld congruence as a notion that makes general sense. But of course, this rests on a viewpoint which may not be widely shared, and which I wouldn't know how to argue for. Also, it now occurs to me, the basic idea of (P4) (the possible worlds definition of our distance comparison predicate) could be achieved in a more complicated way without crossworld congruence, by using a notion of sameness of distance-ratios across worlds (where each ratio is of distances in the same world). So I don't think that my further problem is of much importance: if a relationalist is willing to invoke possible worlds and construe

them in Lewis' hyper-realist manner, then I concede that the problem of quantities is soluble. (That still leaves the argument from field theory for the relationalist to deal with.)

3   My main point in discussing possible worlds was to point out that whereas a Lewis-like conception of possible worlds might solve the problem of quantities, a more moderate conception, such as Stalnaker's would not. But David Lewis has pointed out to me that there is a move that Stalnaker might make which might appear to solve the problem. Besides introducing 'ersatz possible worlds' (as Lewis calls them), why not also introduce 'ersatz possible individuals'? Ersatz possible worlds are (in Stalnaker's terms) 'ways the world might be or have been', that is, they are maximally determinate properties that the whole universe has or might have or might have had; similarly, ersatz possible individuals are 'ways an individual might be or have been', that is, they are maximally determinate properties of a sort that an individual in the universe might have or have had. Once we have such ersatz individuals, why not introduce primitive congruence and betweenness relations Cong_e and Bet_e that apply to ersatz point-particles? (Since 'ersatz point-particles' are properties, Cong_e and Bet_e are higher order predicates, i.e. predicates that apply to properties.) If we do this, we can imitate (P4) as follows:

> (P4e) There are ersatz particles x', y', z', w' and u' in an ersatz world P in which the geometry of Newtonian space-time holds, and in which $u'\text{Bet}_e x'y'$, $x'u'\text{Cong}_e u'y'$, $u'y'\text{Cong}_e z'w'$, $x'y'\text{Cong}_e x^P y^P$, and $z'w'\text{Cong}_e z^P w^P$.

Here x^P is also an ersatz particle, one instantiated in the real world by the genuine particle x; similarly for y^P, z^P and w^P. (We need to distinguish x^P from x: x is a particle, but the 'ersatz particle' corresponding to it, and over which Cong_e and Bet_e are defined, is a property, a 'way x might be' and indeed a way x actually is.) (P4e) is in a sense technically satisfactory. As Lewis noted when he raised the point, though, it really doesn't help the relationalist cause, for it is really just a version of the 'higher order monadicism' already discussed and dismissed in section 2 of this essay.[1]

---

[1] It differs from the version of higher order monadicism originally introduced only in predicating betweenness and congruence of *maximally determinate* properties of a certain sort instead of predicating them of *pure spatial* properties; and it is hard to see how that difference can be of any significance.

7

Realism, Mathematics and Modality*

I advocate a form of anti-realist (or anti-platonist) view about mathematics – one in which modality plays a role, though a rather limited role; and where the modality involved is a modality of the least controversial kind. What my anti-realism involves is a disbelief in mathematics. Or at least, it involves a disbelief in mathematics if mathematics is taken at face value; coupled with a lack of commitment to (and lack of much interest in) the programme of finding a non-face-value interpretation of mathematics on which on the mathematics becomes more believable.

The main goal of this essay is not to defend anti-realism about mathematics – I have done that elsewhere – but to defend the idea that modality has only a very limited role to play in the philosophy of mathematics. In particular, I will be arguing that there are serious difficulties with the idea that talk of possibility can be used as a general surrogate for ontology in mathematics. My views will thus be in opposition to the central theme of Hilary Putnam's work in the philosophy of mathematics from the time of 'Mathematics without

* Much of this essay was prepared for a conference on realism at Northwestern University in 1984; I expanded it for a volume in honour of Hilary Putnam, but because of its length it could not be published there. But I would still like to dedicate this paper to Hilary. Ever since I first read one of his papers, as an undergraduate in 1966, his work has had an enormous impact on me, not only for its specific content but as a model of how philosophy ought to be done. My views on such subjects as scientific realism, scientific methodology, philosophy of mind and philosophy of mathematics probably owe more to him than to anyone else; and despite much doctrinal disagreement with him, I think that no one surpasses him in raising the right questions and correctly locating their philosophical importance.

I have done considerable rewriting of the essay for this volume. Section 2 has been considerably expanded (and section 3 slightly shortened). Section 1 has been greatly shortened, but an expanded version of the material omitted from it has been added as a new section (section 4), and this has led to some significant changes in what is now section 5. I have also corrected the end of what is now section 7.

foundation'; but my view of the role that modality does play in the philosophy of mathematics has considerable affinities to the views expressed in his earlier essay 'The thesis that mathematics is logic'.[1] But it would be well to begin with just a few remarks about what I take the anti-realist position to be. (There will be a bit more on this at the end of the essay, when I consider Putnam's views on realism.)

1 Realism and Truth

I will use the terms 'mathematical realism' and 'platonism' for the view that there are mathematical entities and that they are in no way mind dependent or language dependent. On this usage of 'realism', one form of anti-realism about mathematics might be called mathematical idealism: it is the view that mathematical entities exist, but are somehow mind-dependent or language-dependent. But I find the idea that mathematical entities are mind-dependent or language-dependent rather obscure, and am not tempted to take this way out in opposing mathematical realism. In what follows, I will simply ignore mathematical idealism as an option.[2]

The form my anti-realism takes, then, will be the denial that there are any mathematical entities (numbers, functions, sets and so forth). Since mathematics, taken at face value, postulates the existence of such entities, this means that an anti-realist of the sort I want to be cannot literally believe mathematics (at least when taken at face value).

I have just expressed my mathematical anti-realism by saying that I do not literally believe standard mathematics. In so formulating it, I made no mention of truth. Could I also have expressed my mathematical anti-realism by saying that I do not believe that standard mathematics (taken at face value) is true? I suppose so, *on a certain understanding of 'true'*.

One way to understand the notion of truth is as 'disquotational'. On this reading, a sentence of the form

    '. . .' is true

is to be understood as cognitively equivalent (equivalent by logic plus the meaning of 'true') to the sentence appearing in the blanks, the

---

[1] On some of the affinities and differences between my views and his views in that early paper, see the preface of my 1980, and essay 3 pp. 82–3 and 113–15.

[2] I take mathematical idealism to be not only too obscure to assert, but also too obscure to deny. It may well be that my own view could be redescribed as a view according to which mathematical entities exist but are mind- or language-dependent: this redescription will be appealing to those who like to think of fictional entities as genuine existents created by the authors of the fiction in question.

sentence of which truth was predicated. (Some qualifications are needed: among other things, we must assume the sentence to be substituted in contains no demonstratives or indexicals or the like, and we must assume that it is not one of the 'ungrounded' sentences like those that give rise to the semantic paradoxes.)[3] Now, if one understands truth in this disquotational way, then I have no objections to saying that a mathematical realist must believe that mathematics is true. But I do *not* want to require of a mathematical realist that he or she believe that mathematical sentences are true in some more loaded sense of 'true' – perhaps the sort of sense that has been called 'correspondence truth'. For if the belief that mathematics is true in some 'correspondence' sense leads to difficulties, perhaps the difficulties arise not from belief in mathematics but from belief in the notion of correspondence truth.

To be sure, to deny the legitimacy of a notion of correspondence truth is in a certain sense to adopt a form of anti-realism: it is to adopt an anti-realist attitude about *part of semantic theory* (the part of semantic theory that deals with the alleged property of correspondence truth). Indeed, denying the legitimacy of talk of correspondence truth goes naturally with an anti-realism *about part of psychological theorizing*, i.e., about that part of psychological theorizing that invokes the notion of the truth-theoretic content of psychological states (and utilizes that notion in a way that can't be understood disquotationally). But many people think that denying the legitimacy of a concept of correspondence truth amounts to adopting an anti-realist view about other subject matters as well – e.g., adopting an anti-realist view about the external world, or about science or about mathematics. Well, there is nothing to *prevent* one from so using the terms 'realism' and 'anti-realism', but I do not think that that is a useful way to put matters. For most forms of anti-realism about the external world, or about science, or mathematics, involve either the rejection of some of the assertions about the external world or science or mathematics that have previously been accepted, or else involve some sort of controversial reducibility claim (e.g., of the external world to human experiences) or quasi-reducibility claim (e.g., the claim that theories which agree in what they say about human experiences must be cognitively equivalent). Giving up the idea of correspondence truth does not require us to do any of these things. It allows us to accept these disciplines as fully believable as they stand, and indeed as entirely true in the disquotational sense quite independently of facts about human experience or language.

So let us leave issues of correspondence truth aside. And since it is easy to confuse disquotational truth and correspondence truth, I will

---

[3] See the discussion in fn. 17.

try to avoid talking of disquotational truth too: I will simply say that a realist is someone who believes mathematics and an anti-realist is someone who disbelieves mathematics.[4]

2 A Motivation for Mathematical Anti-realism

I have adopted a relatively weak understanding of what is involved in 'realism': I have denied that a 'mathematical realist' need be committed to a correspondence theory of truth for mathematical sentences. Nonetheless, I believe that even on this weak construal of realism, we should not be realists about mathematics. It is not my purpose in this essay to argue in any detail for this belief; nevertheless, I will mention one of the reasons why I favour an anti-realist conception of mathematics. Essentially it is the familiar epistemological reason: a realist view of mathematics involves the postulation of a large variety of aphysical entities – entities that exist outside of space-time and bear no causal relations to us or anything we can observe – and there just don't seem to be any mechanisms that could explain how the existence and properties of such entities could be known.

The problem can be put without use of the term of art 'knows', and also without talk of truth (though talk of disquotational truth enables us to give a more snappy formulation of it).[5] The mathematical realist believes that his or her own states of mathematical belief, and those of most members of the mathematical community, are to a large extent disquotationally true. This means that those belief states are highly correlated with the mathematical facts: more precisely (and put without talk of truth or facts), that for most mathematical sentences that you substitute for 'p', the following holds:

(1) If mathematicians accept 'p' then p.

Indeed, if various restrictions on the type of mathematical sentence substituted for 'p' are imposed, the converse schema also holds for the most part: there are certain types of mathematical facts which most mathematicians know. Now, the fact that these schemata hold for the most part is surely a fact that requires explanation: we need an explanation of how it can have come about that mathematicians' belief states and utterances so well reflect the mathematical facts. But there seems *prima facie* to be a difficulty in principle in explaining the

---

[4] Recall that mathematical idealism is being left aside.

[5] The view that there can be no conceivable epistemological case against platonism that doesn't rely on a heavy duty notion of truth is curiously widespread. See for instance Tait (1986).

regularity. The problem arises in part from the fact that mathematical entities, as the platonist conceives them, do not causally interact with mathematicians, or indeed with anything else. This means that we cannot explain the mathematicians' beliefs and utterances on the basis of the mathematical facts being causally involved in the production of those beliefs and utterances; or on the basis of the beliefs and utterances causally producing the mathematical facts; or on the basis of some common cause producing both. Perhaps then some sort of *non-causal* explanation of the correlation is possible? Perhaps; but it is very hard to see what this supposed non-causal explanation could be. Recall that on the usual platonist picture, mathematical objects are supposed to be mind- and language-independent; they are supposed to bear no spatio-temporal relations to anything, etc. The problem is that the claims that the platonist makes about mathematical objects appear to rule out any reasonable strategy for explaining the systematic correlation in question.

I do not mean to suggest that the platonist can do nothing toward explaining the general regularity given by (1) and its partial converse. For as mathematics has become more and more deductively systematized, the truth of mathematics has become reduced to the truth of a smaller and smaller set of basic axioms; so we could explain the fact that the mathematicians' beliefs tend to be true by the fact that they have been logically deduced from axioms, if we could just explain the fact that what mathematicians take as axioms tend to be true. That is, what needs explanation is only the fact that the following holds for all (or most) sentences 'p':

(2) If most mathematicians accept 'p' as an axiom, then p.

(Note that what is to be explained is not each instance of (2), but the *general* fact that *all* instances of (2) hold. Note also that this general fact isn't just a fact about those sentences that are taken as axioms. Rather, it is a fact about mathematical sentences generally: namely, the fact that for any mathematical sentence, the disjunction

Either p, or mathematicians don't take 'p' as an axiom

is true.)

Not only can the platonist reduce the problem of explaining the general regularity that instances of (1) hold to the general regularity that instances of (2) hold; he or she can also go *some* way toward explaining the latter regularity. For the truth of all instances of (2) entails the truth of all instances of

(3) If mathematicians accept 'p' as an axiom, then 'p' is logically consistent with everything else they take as axioms;

and (3) does not seem resistant to explanation. For instance, part of the explanation of (3) is surely that the mathematical community has over time managed to weed out the inconsistencies in earlier mathematics – in Newton's beliefs on calculus, in Fourier's on functions, in Cantor's on sets, and so forth. (I think in fact that this is a *large* part of the explanation of (3); this is connected to the fact that a large part of the *reason* most of us believe that modern set theory is consistent is the thought that if it weren't consistent someone would have probably discovered an inconsistency in it by now.) Since (3) is a corollary of (2), an explanation of (3) goes some way toward explaining (2).

These reductions of the problem, though, do not constitute a complete solution of it. That should be obvious: no one would think that we had explained why a person had beliefs about physics that were generally true by pointing out that the person had derived his or her beliefs rigorously from a set of other beliefs that he or she took as axioms, and had subjected the consistency of the axioms to rigorous checking. For there is a big gap between the consistency of an axiomatic theory and its truth. In the case of physics we can presumably fill this gap at least in sketch: we can sketch the route whereby the assumed properties of, say, the electromagnetic field lead to various observable physical phenomena, and thereby affect our perceptual beliefs, and thereby indirectly affect our beliefs about the electromagnetic field. But nothing remotely analogous to this seems possible in the case of mathematics.

It seems to me that this raises a serious epistemological problem for believing in mathematical entities. For if a platonist were to grant that it is impossible in principle to give a satisfactory explanation of the general fact (2), he or she would be left with two unpalatable alternatives: (a) denying that (2) is a fact, or (b) saying that it is simply a brute fact that needs no explanation. But to maintain a class of beliefs while holding the meta-belief that most of those beliefs are false seems plainly unsatisfactory, so we must certainly reject (a). And (b) seems pretty dubious too: there is nothing wrong with supposing that some facts about mathematical entities are just brute facts, but to accept that facts about the relation between mathematical entities and human beings are brute and inexplicable is another matter entirely. I conclude that unless a platonist can make it plausible that it is in principle possible to provide an explanation of the assumed fact (2), then platonism has a serious problem.

The problem I have been sketching is of course a reformulation of the problem made famous by Benacerraf in his paper 'Mathematical truth'. Benacerraf formulated the problem in such a way that it depended on a causal theory of knowledge. The present formulation does not depend on *any* theory of knowledge in the sense in which the causal theory is a theory of knowledge: that is, it does not depend on any assumption about necessary and sufficient conditions for knowledge. Instead, it depends on the idea that we should view with suspicion any claim to know facts about a certain domain if we believe it impossible in principle to explain the reliability of our beliefs about that domain.

In his recent book, David Lewis has adopted a somewhat similar formulation of the Benacerrafian challenge (1986, pp. 111–12); but he holds that it does not pose a genuine problem in the mathematical case, because all facts about the realm of mathematical entities hold necessarily. More fully, Lewis' idea is that we do need – and do have, at least in outline – an explanation of the reliable correlation between the facts about electrons and our 'electron' beliefs (i.e., the beliefs we would express using the word 'electron'); or, as he puts it, we need and have an account (in this case a causal account) of the way in which 'electron' beliefs counterfactually depend on the existence and nature of electrons. But it is only because the existence and nature of electrons is contingent that it makes sense to ask for an explanation of the counterfactual dependence of 'electron' beliefs on the existence and nature of electrons. In Lewis' words:

nothing can depend counterfactually on non-contingent matters. For instance nothing can depend counterfactually on what mathematical objects there are. . . Nothing sensible can be said about how our opinions would be different if there were no number seventeen. (p. 111)

Consequently, since mathematics consists entirely of necessary truths, there can be no sensible problem of explaining why it is that our mathematical beliefs are a reliable indicator of the mathematical facts.

Although I do not want to discuss the Benacerrafian challenge in any detail in this essay, I would like to discuss this particular reply to it. The reply seems at first blush to have a great deal of plausibility; but I think that there are at least four reasons why it fails to undercut the epistemological challenge.

Point one: the premise that all facts about the mathematical realm hold necessarily is false. Let's grant that all facts *purely* about the mathematical realm hold necessarily; that still leaves facts about the mathematical and non-mathematical realms jointly, like such facts (or purported facts) as

(A)   2 = the number of planets closer than the Earth to the Sun;

(B)   for some natural number n there is a function that maps the natural numbers less than n onto the set of all particles of matter;

(C)   surrounding each point of physical space-time there is an open region for which there is a 1–1 differentiable mapping of that region onto an open subset of R^4 (the space of quadruples of real numbers);

(D)   there is a differentiable function ψ from points of space to real numbers such that the gradient of ψ gives the gravitational force on any object per unit mass of that object.

These facts are, by almost anyone's standards, contingent; but they are partly about the mathematical realm, so even if what Lewis says about the explanation of the correlation between the *pure* mathematical facts and our belief states is correct, the problem of explaining the correlation between these 'mixed mathematical facts' and our belief states remains.

There is an obvious strategy for solving the problem of explaining the reliability of our 'mixed' mathematical beliefs, on the supposition that Lewis is right about the purely mathematical beliefs. The strategy is to divide up the mixed mathematical beliefs into two components, a purely mathematical component and a purely non-mathematical component; where the purely mathematical component is a piece of pure mathematics unmixed by physical theory of any sort, and the purely non-mathematical component involves no reference to mathematical entities.[6] The strategy is easy to illustrate, and to carry out, in example (A): (A) can be 'divided up into' the purely non-mathematical claim

(i) $\exists x \exists y (Px \ \& \ Py \ \& \ x \neq y \ \& \ \forall z (Pz \supset z=x \ v \ z=y))$,

(where 'Px' abbreviates 'x is a planet closer than the Earth to the Sun') and the purely mathematical claim

(ii) $[2 =$ the number of u such that Pu] if and only if
     $[\exists x \exists y (Px \ \& \ Py \ \& \ x \neq y \ \& \ \forall z (Pz \supset z=x \ v \ z=y))].$[7]

Because of the fact that (A) is a consequence of (i) and (ii), there can be no problem about the reliability of our belief in (A) unless there is a problem about the reliability of our belief in (i) or (ii). But (i) is a purely non-mathematical belief, and thus explaining its reliability is presumably non-problematic in principle; and (ii) is a purely mathematical

---

[6] Observe that I do not say that the purely mathematical part involves no reference to non-mathematical entities. That would be too strong: it would rule out such purely mathematical theories as set theory with urelements. The important feature of the purely mathematical, for the purposes of the argument in the text, is that it can be regarded with some plausibility as 'necessarily true'.

[7] More precisely, the purely mathematical belief (ii) is such that modulo (ii), (A) is equivalent to the purely non-mathematical belief (i). That is what I really mean by my slightly inaccurate talk of 'dividing up': it is not essential to suppose (as the 'dividing up' talk suggests) that the content of (ii) is part of the content of (A).

belief, and we are assuming for the moment that Lewis' argument works for beliefs of that sort. So 'mixed' statements like (A) present no more of an epistemological problem than do purely mathematical statements; and the same holds for any other 'mixed' statement that can be decomposed into purely mathematical and purely non-mathematical components.

Unfortunately, however, the task of splitting up mixed statements into purely mathematical and purely non-mathematical components is a highly non-trivial one: it is done easily in case (A), but it isn't at all clear how to do it in cases (B) – (D) (at least without introducing some controversial devices). Indeed, as we shall see later (section 5), *the task of splitting up all such assertions into two components is precisely the same as the task of showing that mathematics is dispensable in the empirical sciences*; i.e., the task of showing that in any application of a mixed assertion like (B) or (C) or (D), a purely non-mathematical assertion could take its place. Certainly, then, no one doubtful of the possibility of carrying out the nominalist programme of showing the dispensability of all reference to mathematical entities in science could consistently advocate the strategy outlined in the previous paragraph for solving the problem of explaining our reliable access to the 'realm of mathematical facts' (even if we grant Lewis' claim about the triviality of the problem of explaining our reliable access to the *purely* mathematical facts).

Point two: even with regard to the purely mathematical facts, Lewis' response *at least by itself* is too easy. To explain what I mean, let's ask in what sense mathematical facts (assuming for the moment that there are such) *are* necessary. They are not logically necessary, nor do they (at least on the face of it) reduce to logically necessary truths by definition. (And Lewis clearly did not mean to depend in this passage on the logicist programme of carrying out an acceptable reconstrual of mathematics into pure logic.) They are of course *mathematically* necessary in the sense that they follow from basic laws of mathematics. Similarly, the existence of electrons is presumably *physically* necessary, i.e., follows from basic physical laws. But Lewis does not think that the epistemological problem of explaining how our 'electron' beliefs can reliably indicate the existence of electrons is really a pseudo problem, just because the existence of electrons is physically necessary; so why should the fact that the existence of numbers is mathematically necessary show that the corresponding epistemological problem about numbers is a pseudo-problem?

One might try to answer this by saying that mathematical necessity is *absolute* necessity while physical necessity is necessity only of a restricted sort; or that mathematically necessary claims are *metaphysically* necessary, while physically necessary claims are not. But it is hard to see how to give any content to this which is of any help. In the first place, it is worth noting that there are perfectly good senses in which mathematical necessity is *not* unrestricted necessity: e.g., it is not logical necessity. What then is the content of the distinction between 'absolute' or 'metaphysical' necessity and necessities of lesser sorts, and what are the grounds for saying that mathematical necessity is of the former sort but that physical necessity is not? One could just *stipulate* that the term 'absolute necessity' is to cover mathematical necessity but not physical necessity. But then the question of the previous paragraph becomes: what are the grounds for thinking that it is absolute necessity so defined rather than, say, logical necessity or physical necessity, that is epistemologically relevant? Alternatively, one could try to somehow build it into the definition of 'absolute necessity' that the epistemological problem can't arise for absolutely necessary truths; but then the question is, why suppose that mathematical claims unlike claims of basic physics are 'absolutely necessary' in this sense. Clearly the introduction of talk of 'absolute necessity' is of no help: on either of the two conventions for using this phrase, we need an argument that the epistemological problem doesn't arise for claims that are mathematically necessary, i.e., that follow from the accepted mathematical axioms; in particular, we need an argument that the epistemological problem does not arise for the mathematical axioms themselves. Unless an answer to this is given, one suspects that the claim that mathematical necessity is 'absolute necessity' simply amounts to the decision not to take the problem of explaining the reliability of our mathematical beliefs seriously; if so, it cannot be used to justify that decision.

The point I am making can be reinforced by noting how easy it is to invoke 'metaphysical necessity' to defend metaphysical beliefs against epistemological challenge. If a believer in God is embarrassed by an inability to explain the fact (i.e., what he or she takes to be a fact) that his or her belief in God and beliefs about God are fairly reliable, the solution is simple: declare the problem a pseudo-problem on the grounds that God is a necessary being and has all his properties necessarily. If a believer in possible worlds as Lewis conceives of them – that is, possible worlds more or less like the real world but bearing no spatio-temporal relations to the real world – is embarrassed by inability to explain the alleged reliability of our beliefs about these other-worldly entities, the solution is again simple: declare the problem a pseudo-problem, on analogous grounds. Lewis himself, of course, employs this tactic in the possible worlds case (though not in the theological case): indeed, in the passage I quoted Lewis' main concern was to undermine epistemological objections to his conception of possible worlds. I suspect that many readers of this essay will feel that the application of the tactic to the possible worlds case (possible worlds construed as Lewis construed them, not simply as mathematical entities), and to the theological case, is a cheat; but that in application to the mathematical case the tactic is reasonable. Maybe so, but if so some justification is needed for this choice of which cases to regard the tactic as legitimate and which case not. I myself incline to the more even-handed view that the tactic is unreasonable in all three cases.

In the next section I will give a diagnosis of the common view that there is a sense of necessity that is akin to logical necessity in a way in which physical necessity is not, in which mathematics is, if true, necessary. If this diagnosis is correct, I think it will further undermine the idea that the problem of explaining the reliability of our beliefs about the mathematical realm, given the assumption that there is such a realm, is a pseudo-problem.[8]

On points three and four, I'll be briefer. Point three is that Lewis is assuming a controversial connection between counterfactuals and necessity, and that even those who think that there is some sort of 'absolute necessity' to mathematics may find counter-mathematical conditionals perfectly intelligible in certain contexts. It is doubtless true that nothing sensible can be said about how things would be different if there were no number 17; that is largely because the antecedent of this counterfactual gives us no hints as to what alternative mathematics is to be regarded as true in the counterfactual situation in question.[9] If one changes the example to 'nothing sensible can be said about how things would be different if the axiom of choice were false', it seems wrong (assuming platonism – see footnote 9): if the axiom of choice were false, the

[8] In this discussion I have avoided taking a stand on whether even logical necessity should be viewed as 'absolute' necessity. One view, to which I am attracted, is to reject the whole notion of 'absolute' necessity as unintelligible. Another view, also with some attractions, regards the notion as intelligible but regards the only things that are absolutely necessary as logic and matters of definition; in particular (as Hume and Kant held) there can be no 'absolutely necessary' entities, and so mathematics (taken at face value) cannot be absolutely necessary since it implies the existence of mathematical entities of various sorts.

Whichever of these two views one prefers, the fact that mathematics involves existential commitment in a way that logic doesn't makes for a sharp distinction between mathematical necessity and logical necessity. The view that these two sorts of necessity are closely akin cries out for explanation. The only candidate I know of for explaining this is the one I offer in my 'hygienic explanation of mathematical necessity' in section 3 below.

[9] Actually, a fictionalist can explain the unintelligibility of the counterfactual in a different manner: there *is* no number 17, so the question is like asking how things would be different if there were no Santa Claus. But let us assume platonism for the moment, since we are discussing an epistemological problem for the platonist.

cardinals wouldn't be linearly ordered, the Banach-Tarski theorem would fail and so forth.

But what should a platonist say about what *the opinions of mathematicians* would be like if the axiom of choice failed? That is a hard question, not because it seems unintelligible, but because it poses a dilemma. There is a strong case to be made for saying that if the axiom were false, mathematicians' opinions would be just as they are now: after all, sets don't causally interact with mathematicians, nor do they stand in any other sorts of relations to mathematicians that would seem to license the conclusion that the mathematicians' opinions would be different. But if we are platonists, we will *want* to say that if the axiom were false, then mathematicians would believe it to be false; that's what we seem to need to say if our present opinion that the axiom is true is to have epistemological value. The epistemological problem, if put in terms of counterfactuals, is to figure out how to say what we (in our platonist moods) want to say, rather than what we seem forced to say by the causal inertness, mind-independence etc., of the mathematical objects.

My fourth and final point is this: there was no obvious reason to put the challenge of explaining the reliability of our mathematical beliefs in modal or counterfactual terms in the first place. Indeed, the phenomenon that our beliefs about (say) electrons are reliable is not *simply* that our 'electron' beliefs counterfactually depend on the facts about electrons: it is that our beliefs depend on the facts about electrons *in such a way that* the correlation of our believing the sentence 'p' and its being the case that p would be maintained given a variation in the facts about electrons. It is *this type of* counterfactual dependence that needs explaining, not counterfactual dependence by itself. But now, if the intelligibility of talk of 'varying the facts' is challenged in the mathematical case, it can easily be dropped without much loss to the problem: there is still the problem of explaining the *actual* correlation between our believing 'p' and its being the case that p. That, in fact, is the way I put the problem at the start of this section.

I have in a very sketchy way tried to make it plausible that if one postulates mathematical entities one is going to have a serious problem explaining the correlation between our mathematical beliefs and the facts about those entities; and I have tried to argue that *to the extent that providing such an explanation appears impossible*, believers in a mathematical realm face a genuine epistemological problem. Note that this way of putting the worry about mathematical realism does not involve the claim that there can be no good reason to believe in mathematical entities. That claim would be quite wrong, I think: there can be strong reasons to believe in mathematical entities, having to do with the apparent indispensability of mathematical entities to important theories outside mathematics; I believe that these reasons can ultimately be rebutted, but it isn't entirely clear how best to rebut some of them. The point, rather, is that however good the reasons *for* believing in mathematical entities, the difficulty in explaining the reliability of those beliefs remains, and it raises a serious epistemological puzzle for those not happy with taking a perception of the realm of mathematical entities as primitive. That epistemological puzzle is a reason against believing in mathematical entities, which has to be weighed against the reasons in favour of mathematical entities. Actually it isn't a simple case of weighing: we have a case of competing arguments; for a satisfactory view to be achieved, we must find a way of disarming one of the competing lines of argument. I do not declare it misguided to try to show that it is possible to explain the reliability of our access to a purported realm of mathematical facts;[10] or to try to show (as Lewis does at least in the case of pure mathematical facts) that the lack of such an explanation should not be upsetting.[11] But I know of no way to do either of these things at all convincingly, and my bet is that the anti-realist programme of undercutting the reasons for literally believing mathematics is more promising. That is all that I shall say about these epistemological matters here.

3 A Version of Mathematical Anti-realism

The form that my own anti-realism takes is very straightforward: it involves no claim that mathematical statements mean anything other than what they appear to mean; instead, it simply claims that there are no mathematical entities, and hence that mathematical statements which assert that there are such entities are not true. This means that most if not all atomic statements in mathematics are untrue (if taken at face value, as I prefer to take them), as are all statements whose main connective is an existential quantifier purporting to range over mathematical entities; while statements whose main connective is a universal quantifier over mathematical entities are *vacuously* true, even if they conflict with established mathematical theories. Obviously I am not proposing that we replace the usual mathematics by a new mathematics that rejects all existential assertions and accepts all universal ones: such a body of assertions would be of no mathematical interest. What I am

---

[10] For an attempt to do this, see Maddy (1980).

[11] For another attempt to dismiss the sort of epistemological problems involving mathematics that I have been considering, see Wright (1983). I have argued against this attempt in essay 5.

asserting, rather, is that what is of mathematical interest and what is true do not coincide. As van Fraassen (1980) put a rather similar doctrine about science, I am proposing that mathematics 'does not have to be true to be good'. And here 'true' is being used in a purely disquotational sense. The notion of truth could be avoided here by simply saying that the assertion 'There are infinitely many prime numbers' can be 'mathematically good' even if there are no numbers at all (hence certainly not infinitely many prime numbers); and that the same holds for many other mathematical assertions.

If mathematics needn't be true (even disquotationally true) to be good, then what condition must it satisfy? One goal that has occasionally been suggested as an alternative to truth is consistency. But in essay 2 I argued that this suggestion is untenable: a consistent mathematical theory could be grossly inadequate, for it could imply false conclusions about the physical world. (Consider as an example the variant of set theory that I discussed in that essay.) Instead, I proposed as my alternative to truth a slight strengthening of consistency, which I called *conservativeness*. To say that a theory is conservative is to say that it is consistent with every internally consistent theory that is 'purely about the physical world' in that it involves no reference to mathematical objects.[12]

It might be objected that there are many conservative theories that a mathematician would have no interest in, and therefore that 'mathematical goodness' can't be mere conservativeness. But by the same token, there are many mathematical theories that any mathematical realist would regard as *true* but which no mathematician would have any interest in – for instance, the theory that asserts only that there are at least two mathematical objects. Consequently, 'mathematical goodness' can't consist only of truth either. Clearly there are many factors that go into deciding whether a given mathematical theory is a piece of good mathematics – e.g., richness in consequences, relevance to prior work in mathematics and in science, elegance and originality, to name just a few. But for the mathematical realist, *one* such factor is truth. What I am proposing is that the anti-realist should put conservativeness *in place of this one factor*; I do not deny that the other factors should be operative as well.

What is the relation between conservativeness and truth? First, conservativeness does not require truth. This should be clear merely from the definition of conservativeness; but in addition, it is worth noting that conflicting mathematical theories (e.g., set theory with the

---

[12] 'Consistent' here means (roughly) 'semantically consistent', i.e., 'satisfiable' – cf. essay 4. The reason for the 'roughly' is given in essay 3, and will be repeated shortly.

axiom of choice and set theory with its negation, or set theory with the continuum hypothesis and set theory with *its* negation) can both be conservative. So the claim that mathematical theories are conservative certainly does not entail that they are true (even in the disquotational sense of 'true') and, consequently, conservativeness looks like a goal that the mathematical anti-realist can consistently advocate.

Second, truth doesn't require conservativeness: for true *empirical* theories are *obviously* not conservative. The conservativeness of mathematics is, then, a noteworthy feature of mathematics, a feature that does not follow from the alleged fact that mathematics is true. But I think that most mathematical realists have implicitly recognized the conservativeness of mathematics: for they have held that good mathematical theories are not only true but necessarily true; and a clear part of the content of this (the *only* clear part, I think) is that mathematics is conservative. (Mathematical realists like Quine who want to scotch all talk of mathematical necessity presumably need to recognize the conservativeness of mathematics; though doing so would force considerable qualifications on the Quinean view that mathematics is continuous with the rest of science and subject to revision in the same way that science is subject to revision.)[13]

So the relation between conservativeness and truth, necessary truth and consistency can be diagrammed as follows:

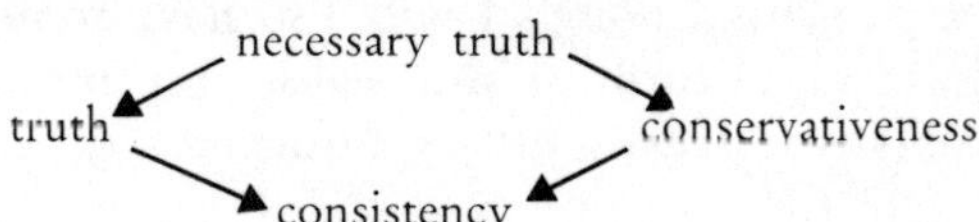

Conservativeness might loosely be thought of as 'necessary truth without the truth'. As I have remarked, I think that the notion of necessary truth which a realist typically appeals to is quite an obscure one, but I think that the only *clear* difference between a conservative theory and a necessarily true one is that the conservative theory need not be true at all. Perhaps many realists would be content to say that all they ever

---

[13] Qualification, not elimination. In the first place, arguments for expanding mathematics to contain hitherto undreamt of sorts of entities could still be made on empirical grounds, for such expansions needn't be motivated by the old mathematics (plus nominalistic auxiliary premises) leading to incorrect predictions (that the auxiliary premises themselves don't lead to), they could be motivated simply by the old mathematics being incapable of expressing things that need expressing. In the second place, the conservativeness of mathematics doesn't even rule out the possibility of empirically disconfirming mathematics (in the sense of disconfirmation provided by a roughly hypothetico-deductive model), as long as we allow (with Quine) that *logic* can be empirically disconfirmed. However, the conservativeness of mathematics does rule out the possibility of empirically disconfirming mathematics while leaving logic intact.

meant when they called mathematical claims necessarily true was that they were true and that the totality of them constituted a conservative theory. Note, though, that this explication of what it means for mathematics to be necessarily true could not be employed by anyone who wanted to dismiss the epistemological problem raised by mathematics in the manner considered in section 2. If a role of the traditional notion of necessary truth is to perform epistemological magic, then that role is not preserved by the hygienic explication in terms of conservativeness.

The debate between mathematical realism and the form of mathematical anti-realism I advocate, then, is a debate between whether mathematics need be true (or true and conservative, or necessarily true and hence conservative), or whether it need only be conservative. Grounds for favouring the anti-realist position are obvious: it is more theoretically economical; it enables us to avoid viewing the choice between two conservative mathematical theories as a matter of right or wrong about which we could well be making a mistake; and in general it enables us (without appeal to epistemological magic) to avoid the epistemological question touched on in section 2 concerning how we can have any reliable access to the alleged mathematical facts. But of course, there are alternative considerations that favour the realist view. In the rest of this essay I will mostly be discussing whether the invocation of modality will be of any help to the mathematical anti-realist in trying to undercut pro-realist arguments; though I will also have something to say about a slightly less radical form of anti-realism (or anyway, anti-platonism) that has been advocated by Hilary Putnam.

4 Truth and Substitutional Quantification

According to Frege, 'it is applicability alone which elevates arithmetic from a game to the rank of a science.' (1966, p. 187.) Generalizing slightly, one might argue that the only real argument for thinking that mathematics is (disquotationally) true rests upon its usefulness outside mathematics; and indeed I think that this is extremely plausible. But to what extent is the utility of mathematics evidence for its truth? One use of mathematics is to facilitate inferences among purely nominalistic assertions: we see easily that nominalistic assertion A follows from nominalistic theory N plus mathematics, whereas if we don't use mathematics it is hard to see that A does follow from N. Such uses of mathematics are extremely important; but they provide no evidence whatever for the truth of mathematics (even in the disquotational sense of 'true'), since the legitimacy of inferring A from N when one has deduced A from N plus mathematics doesn't depend on the mathematics being true, it requires only that the mathematics be conservative.

But mathematics also seems to be of value in another way: it seems to be *theoretically indispensable*, that is, needed among the premises of important theories, including empirical theories. (This is not ruled out by the conservativeness of mathematics: the conservativeness of mathematics tells you something about how you can use mathematics when you have a nominalistic theory, but the issue here is whether sufficiently powerful nominalistic theories are available.) Unlike van Fraassen, I concede that if mathematics is theoretically indispensable, then that is a powerful argument for its truth. So as a mathematical anti-realist, I want to deny that mathematics really is theoretically indispensable.

In *Science without Numbers* I defended the view that mathematics is not theoretically indispensable *in science*: scientific theories can be nominalistically reformulated,[14] I argued, and when they are, the usual platonistic formulations of science can be justified (as legitimate to use even though not strictly true) by using the conservativeness of mathematics. It is not my intention to review or defend these claims here. (Defence of them against one sort of widespread quasi-technical objection can be found in essay 4. Essay 6 is also relevant to this objection, and more directly relevant to another objection concerning my reliance on a substantivalist conception of space-time.) But I do want to stress one thing, and that is that the reformulation of Newtonian physics that I endorsed in illustrating my thesis made no use of modal operators. One question that I want to discuss later in this essay (sections 6 and 7) is whether I mightn't have dealt with the indispensability arguments for the truth of mathematics more easily (and in a way that may be applicable to a wider class of physical theories) if I had allowed the use of modality in reformulating physics.

But before I get into that issue, I want to consider briefly another kind of indispensability argument for mathematical entities: I want to consider the argument that mathematical entities are needed to account for metalogical notions like (disquotational) truth, consistency and conservativeness.

Let's discuss disquotational truth first; and in our initial discussion, let us ignore ontological scruples about mathematical entities, so that we can see what the most plausible way to treat disquotational truth platonistically is before we discuss what the most plausible nominalistic treatment is.

[14] Indeed, the nominalistic formulations that I proposed are attractive independently of ontological scruples. This point was emphasized in Field (1980), but stated more clearly in section 5 of essay 6.

Disquotational truth is a notion that applies primarily to sentences of one's own language. (It can of course be extended to other languages if one has either a particular translation in mind or else a notion of correct translation.) Since '". . ." is (disquotationally) true' is supposed to be cognitively equivalent to '. . .', there would be no theoretical importance in having a notion of disquotational truth if it could only be attached to single quoted sentences, since it would always be eliminable without loss of cognitive value. But of course the notion of disquotational truth is applicable in quantificational constructions too; and as Quine (1953, 1970) and Stephen Leeds (1978) have pointed out, this can give the notion of disquotational truth a real value in increasing the expressive power of a language: specifically, *a disquotational truth predicate serves as a device of infinite conjunction.*[15]

It seems to me that having a device of infinite conjunction at our disposal is pretty nearly essential. Typically, theories about the physical world are formulated with infinitely many axioms (finitely many separate axioms, plus finitely many axiom schemata each of which has infinitely many axioms as instances). Some sense can be made of what it is to accept such a theory without any means to conjoin these infinitely many axioms: one could say that to accept the theory is to accept each of the separate axioms and be disposed to accept each instance of the axiom schemata. (This is a slightly weak sense of accepting a theory, but let's let that pass.) But the idea of *rejecting* such a theory is more problematic, given that one can reject a theory on the grounds that it has unacceptable consequences without knowing which part of the theory ought to go. If a theory is finitely axiomatized, then, we must say not that we reject some axiom, but that we reject the conjunction of the axioms; but if the theory is not finitely axiomatized, then there *is* no conjunction of the axioms in any straightforward sense. We need to introduce some sort of device of infinite conjunction. One possibility is to introduce a disquotational truth predicate and semantically ascend, saying 'not every axiom of this theory is true.' It seems to me that if for some reason we had to disallow a disquotational truth predicate, we should find some means of getting the infinite conjunctions without one.

Before asking whether a disquotational truth predicate is available to a mathematical anti-realist, let's contemplate the idea of getting infinite conjunctions more directly, without semantic ascent. Logicians have

---

[15] Also of finite conjunction where the class of sentences being conjoined is too large to be explicitly given, or where the members that comprise the class are unknown, as in 'Everything Jones ever said is true.' This case, though, is reducible to the infinite case ('Every sentence is such that either Jones never said it or it is true'), and while the reduction may be artificial it will simplify my formulations if I employ it.

studied languages with devices of infinitary conjunction. The languages they have mostly studied (like $L_{\omega_1\omega}$) contain the means for expressing *even highly irregular* infinite conjunctions; we need only a language with the means of expressing the *sufficiently regular* ones expressible by a notion of disquotational truth. It isn't hard to see that such a language doesn't even need to be infinitary in any very strong sense: that is, a finite notation can suffice. Indeed, a first order language that contains 'substitutional quantifiers' (with formulas as substituends) in addition to ordinary quantifiers suffices for most or all of the infinite conjunctions that we need. Such a device of 'substitutional quantification' (viewed simply as highly 'regular' infinite conjunction) can in fact be used to define disquotational truth,[16] if we wish to do so:

S is true if and only if Πp (if S = 'p', then p),

where 'p' takes sentences as substituends.[17] But the substitutional quantifier (or other device of infinite conjunction) also allows us to

[16] I hesitate to use the term 'substitutional quantification', since many view substitutional quantification as defined in terms of objectual quantification over expressions. The name suggests this; and many have wondered how else it could possibly be interpreted. (See for instance van Inwagen (1981).) I take it, though, that the infinitary conjunction line provides the alternative interpretation: '$\Pi F \Pi G$(If John is F then he is F or he is G)' is no more about expressions than are finite conjunctions like 'If John is happy then he is happy or sad, and if John is angry he is angry or sad.' This same analogy to finite conjunctions shows that one cannot argue from the fact that a truth theory for 'substitutional quantifiers' would quantify over sentences to the claim that they are to be interpreted in terms of sentences: this line of reasoning is obviously fallacious in the case of finite conjunctions. (I should add that the temptation to argue in that fashion seems to result from the illusion that a truth theory has an important role to play in an account of how a language is understood.)

[17] One must be careful about *what* sentences are taken as substituends, of course: for the idea of a conjunction of some sentences only makes clear sense if those sentences are understandable prior to the conjunction; this means that the conjunction itself, and sentences constructed from it, shouldn't be allowed as conjuncts. Failure to observe this requirement, or a slight weakening of it to be mentioned shortly, can result in analogues of the semantic paradox (analogues since they employ not notions like truth, but other devices of infinite conjunction). The most obvious way to enforce this requirement is to mimic the Tarski hierarchy of truth predicates: let a level 0 sentence be one containing no device of infinite conjunction; we then allow 1-sentences that contain (regular) infinite conjunctions of 0-sentences; 2-sentences that contain infinite conjunctions of 0- and 1-sentences; and so on. The definition of disquotational truth in terms of substitutional quantification is then typically ambiguous, producing each member of the Tarski hierarchy from the corresponding substitutional quantifier.

Kripke (1975) argues that the Tarski hierarchy is too restrictive for our needs: in effect, it fails to give us all the infinitary conjunctions we want. He suggests a broader, quasi-impredicative, conception of disquotational truth. This conception has an analogue for substitutional quantification, as Kripke observes, and in fact for infinitary conjunction

cont. over page

make infinite conjunctions without semantic ascent. This seems to me a preferable way to proceed: if what we want to do is talk about the non-linguistic world, why bring in sentences if we don't have to?

One policy, then, is to take substitutional quantification or some other device of infinite conjunction as primitive, in the same way that negation or conjunction or ordinary quantifiers are taken as primitive. To say that we take it as primitive means (a) that we do not regard it as getting its meaning from a definition of it in other more basic terms, and (b) that we do not regard it as needing theoretical explication in other more basic terms. I think that both attitudes are plausible. Of course, if we adopt attitude (a) then we must say how the notion does get its meaning. Presumably the answer is that it gets its meaning from the logical laws taken to govern it. (I will not discuss here just what those laws are.) If attitude (a) is adopted, I see no problem with also adopting attitude (b). (There are various notions – e.g., mentalistic notions – whose meaning derives from the theory or predictive practice in which they function, but are often felt (correctly, I think) to require theoretical explication: the point of the explication is not to give the meaning of the notion in question but to show that it is non-basic. But in the case of a logical notion, I see little motivation for requiring such ideological reduction.) In any case, if this is the attitude that we adopt then, as I've said, disquotational truth is both definable and dispensable – dispensable independent of the definition, because we can serve the need for infinite conjunctions without semantic ascent. An alternative attitude, from the platonist point of view almost equivalent, is to take disquotational truth itself as primitive (i.e., undefined and not needing theoretical explication), and as clarified by the disquotation schema

'p' is true if and only if p

(or perhaps the result of attaching a necessity operator to this since truth is disquotational) caged with suitable restrictions so as to avoid semantic paradox. This is perhaps slightly less natural – it may seem a bit odd to think that a predicate of such special entities as sentences needs no theoretical explication, and if so, this will be a reason to

---

more generally: the idea is to allow sentences to contain themselves and things built out of them as conjuncts, but to let the truth value of such quasi-impredicative conjunctions be objectively indeterminate unless it is settlable at some level, so to speak. The definition in the text then defines the Kripkean notion of disquotational truth from Kripkean substitutional quantification.

Ultimately I think it is Kripkean substitutional quantification that we should employ. However, for purposes of this essay, insofar as we need substitutional quantification at all it is only very 'low level' applications of it: we could stick to the lowest two levels of the Tarski hierarchy.

explicate it in terms of infinitary conjunctions if we are to use it at all – but I will not pursue this.

But isn't there another option? Didn't Tarski show how to define or explicate disquotational truth purely in mathematical terms without reliance on substitutional quantification? No, he did not, and indeed he proved that it could not be done. If M is our broadest mathematical theory, we cannot define in M a notion of disquotational truth applicable to arbitrary sentences of M; we can define disquotational truth only for the sentences of M in which the quantifiers are restricted to range over only the members of some set. (Or if you prefer, we can define a predicate 'true in domain D' (where D is a set) for all sentences of M; but truth in D is not disquotational truth.)[18] It is irrelevant that we could define truth for M if we went to a higher order mathematical theory M*, for we still wouldn't be able to define truth for M*; whatever mathematical theory we pick, the full notion of disquotational truth we have when we pick it won't be definable in the theory we've picked. We do, though, seem to understand such a notion, and seem to need it, for instance when we conjecture that our total mathematical theory might not be true.

Tarski's approach to the theory of truth is unduly restrictive; an approach that is more general, in that it allows for a single notion of truth that is applicable to sentences themselves containing 'true', has been given in Kripke 1975. But the idea that disquotational truth is mathematically definable, without infinitary devices like substitutional quantification, is no more supported by Kripke's approach than by Tarski's:[19] if anything, it is *less* supported, because on Kripke's approach even the *recursive* definition of 'true' (for sentences that may contain 'true') is possible only for sentences with quantifiers restricted to some

[18] To be sure, the restriction of the quantifiers is needed only at the stage where Tarski turns a recursive definition into an explicit definition; this means that if our goal is only to *recursively* define 'true', Tarski's definition is adequate, at least as applied to sentences that don't contain 'true'. One might argue that recursively defining a concept is sufficient to give it clarity, even in absence of an explicit definition: don't we understand addition pretty well on the basis of the usual recursion clauses, independently of the trick for making the definition explicit? I think that this is plausible, but derives its plausibility from the legitimacy of a notion of disquotational truth or some other device of infinitary conjunction. (We understand addition from the recursion clauses because they allow us to conjoin the instances, or to know that each of their instances is disquotationally true.) If this is right, then recursive definitions cannot be employed in place of explicit definitions when it is disquotational truth itself that is in question.

[19] Indeed, as long as we assume that truth value gaps do not arise when 'true' is predicated of a sentence that doesn't contain 'true' (or other truth-theoretic terms), then the Tarski indefinability theorem immediately rules out a definition of 'true' (in non-truth-theoretic terms).

set as its domain. What makes Kripke's extension of Tarski valuable is not that he has enabled us to do better at defining disquotational truth, but that his definitions of *restricted approximations to* disquotational truth (approximations which allow 'true' to be applicable to sentences themselves containing 'true') give us a broader conception of how disquotational truth generally should work. The broader conception should be reflected in an axiomatic treatment of the notion of disquotational truth, or (better I think) in an axiomatic treatment of a closely related notion of substitutional quantification or of infinitary conjunction more generally. (See footnote 17.)

I've been arguing that even a platonist should regard disquotational truth not as being *mathematically defined* in the manner of Tarski or Kripke, but as either a primitive notion or (almost the same thing, but probably slightly preferable) as defined from a notion of substitutional quantification or some more powerful form of infinitary conjunction. When we turn to anti-platonism, there is added reason to choose the substitutional quantification/infinitary conjunction line over the primitive disquotational truth line (at least if our anti-platonism is sufficiently radical). The reason is that disquotational truth is a predicate of sentences, and a sufficiently radical anti-platonist will not believe that there are any sentences except those that have been spoken or written; this will create problems for many applications of the notion of disquotational truth. The way to avoid this difficulty, I think, is to get our infinite conjunctions directly, without semantic ascent.[20]

---

[20] Could we get around the problem with using disquotational truth itself as a primitive, by appeal to some kind of modality? For instance, could we employ the idea of *it being possible for there to be a disquotationally true sentence such that* . . . as a surrogate for the idea of *there actually being a disquotationally true sentence such that* . . .? I will not take a stand on the workability of this sort of approach here, but some of my remarks later in the essay, critical of certain other attempts to use modality to avoid ontological commitment, certainly seem at least at first blush to tell against it.

On a different matter: there are several problems that bear looking into about the status of the 'primitive infinite conjunction' approach for the sufficiently radical anti-platonist. The first sort of problem is that we need to get a flexible enough device of infinite conjunction to allow us to make all the infinite conjunctions we need without appeal to uninstantiated expression types. It is easy to see that ordinary substitutional quantification is problematic in this regard. The second problem is that we need to get a reasonable set of procedural rules for operating with the device of infinite conjunction. I think that in the case of ordinary substitutional quantification we can do this, but that there is a chance that we will get problems when we turn to a more flexible kind of infinitary conjunction as needed for the first problem. These are important issues, but they are also complex, and I will not attempt to pursue them here.

5 Modality and Metalogic

The question of whether a mathematical anti-realist can make sense of metalogical notions seems especially pressing in the case of notions like consistency and conservativeness. After all, the anti-realist wants to assert of typical theories in mathematics that they are consistent, and indeed conservative: so if the anti-realist can't make sense of claims of consistency and conservativeness, it would seem that anti-realism about mathematics is a self-defeating doctrine.

And it may seem to be unclear how a mathematical anti-realist could make sense of consistency and conservativeness. For consistency is usually defined as *having a model*, where a model is a set of a certain sort, and hence a mathematical entity; and since I defined conservativeness in terms of consistency, it will be defined model-theoretically as well. If the anti-realist were to accept these definitions, then there could be no non-vacuous distinction between consistent theories and inconsistent ones, or between conservative ones and non-conservative ones, so that the anti-realist view *would* be self-defeating.

Consequently, the anti-realist should reject the idea that consistency and conservativeness are to be defined in the usual model-theoretic fashion, just as he or she should reject the idea that truth is to be defined in the usual Tarskian-Kripkean fashion. But this is not a defect in the anti-realist position, I think, for it seems to me that a very strong case can be made that *even the platonist* should reject the model-theoretic definitions. I have argued this in section 5 of the introduction, and will not repeat the argument here.

In order to motivate what the anti-realist *should* say, let's go back to the diagram in section 3, illustrating four possible views that someone might take about one's mathematical theory T: that it was true, that it was necessarily true, that it was consistent, and that it was conservative. These are all metalinguistic claims about the theory T. In the case of truth, though, we've seen that what the mathematical realist wants to assert is not really a metalinguistic claim in any ordinary sense: what he or she wants to assert is simply T itself. That is, he or she wants to assert each axiom of T. Indeed, he or she wants to assert each finite conjunction T_i of axioms of T. (This last is a stronger claim, for anyone who doesn't think that acceptance is closed under consequence.) And indeed, if you buy talk of infinite conjunction, he or she wants to assert AX_T, the conjunction of every axiom in T. Any of these conditions is sufficient for realism, I argued in section 1: after all, the existence of mathematical entities follows *from the mathematical theory itself*, not just from the claim that the mathematical theory is *true* in the correspondence sense; so anyone who believes a mathematical theory believes in mathematical entities and consequently faces the usual epistemological difficulties that a believer in mathematical entities has to deal with.

Something similar holds in the case of consistency. A mathematical anti-realist needn't hold that our mathematical theory T has a *property* of consistency in any ordinary sense. Rather, what must be believed can be explained without reference to the theory T at all: what the mathematical anti-realist must believe is that *it is logically consistent that T_i*, for each finite conjunction T_i of axioms of T. Here, 'it is logically consistent that' is an operator that is attached to the sentence T_i; it is *not* a *predicate* of that sentence, in the way that 'true in some model' or 'true in some possible world' are predicates of the sentence. If a device of infinitary conjunction is available, this can of course be expressed in a single sentence. (In either of two ways: either the conjunction of the claims 'it is logically consistent that T_i', or as the claim that it is logically consistent that AX_T where again AX_T is the conjunction of the axioms. These are equivalent if the logic is compact; if it is not compact, the latter is preferable.) The possibility of representing the consistency of an infinitely axiomatized theory in a single sentence, without semantic ascent, depends of course on our having a device for infinite conjunction that does not involve semantic ascent. If we disallow such a device, we can still assert the consistency of T by asserting each claim of form 'it is logically consistent that T_i'; or we can rely on semantic ascent, and say it is logically consistent that T is disquotationally true.[21]

The requirement of necessary truth which most mathematical realists would impose can similarly be given at the object level, provided that we recognize an operator '$\Box_m$' of mathematical necessity (or some more general *a priori* necessity, perhaps 'absolute necessity'), weaker than the purely logical necessity operator '$\Box$'. ('$\Box$', also read 'it is logically true that', is of course equivalent to '$-\Diamond-$', where '$\Diamond$' means 'it is logically consistent that'.) For instead of asserting that T has the property of necessary truth, the necessitarian realist can assert that $\Box_m T_i$ for each finite conjunction T_i; or that $\Box_m AX_T$ if you want to talk in terms of infinite conjunctions.

It is worth repeating that a realist who imposes such a necessity requirement can't make do with a purely logical necessity: the axioms

---

[21] The standard objection to using *correspondence* truth in this connection – viz., that even for an *inconsistent* theory, it is *consistent* that it is true, since the logical words in it could have been used differently – does not apply to disquotational truth, since the latter is itself a logical notion and the instances of the disquotation schema are to be conceived of as logical truths.

of mathematical theories are not logically necessary, for if they were we wouldn't need the mathematics and could settle for logic. As I have said, it is to my mind quite obscure what the notion of mathematical necessity or absolute necessity is supposed to come to,[22] and it is not an idea that I shall have any use for; my point is only that if it is regarded as coherent then it too is best treated in terms of an operator rather than a predicate.

Finally we come to the fourth node of the diagram: conservativeness. If we are to avoid any appeal to infinitary conjunction, then to assert that T is conservative is to assert each sentence of the form:

If it is consistent that B then it is consistent that B^* & T_i,

where B is any sentence without mathematical vocabulary, B^* is the result of restricting the variables of B to non-mathematical entities, and again T_i is any conjunction of axioms of the theory T. (See point 2 in the postscript to essay 3 for some discussion of the adequacy of this formulation.) Again, we can conjoin these instances into a single sentence if we so choose, with (two applications of) infinitary conjunction. The result would be a modal statement C_T (using only a notion of *logical* possibility) which more or less follows from $\Box_m AX_T$ (on any reasonable stipulation as to what '$\Box_m$' means),[23] which implies $\Diamond AX_T$, and which neither entails nor is entailed by AX_T. This modal statement C_T, the modal statement of the conservativeness of T, thus fits into the diagram in the appropriate way.

The upshot then is that our entire diagram can be transcribed on to the object level. The transcribed version of the truth of T – viz., T itself, or equivalently AX_T – is problematic since it involves a commitment to mathematical entities. The transcribed version of the necessary truth of T is doubly problematic since it involves not only mathematical entities but also a rather obscure notion of *mathematical* as opposed to purely logical necessity. But the transcribed versions of consistency and conservativeness are problematic in neither of these two ways.

---

[22] This is not to deny that clear explications are possible: for instance, I have given a 'hygienic' explication of mathematical necessity earlier (second last paragraph of section 3). But in the hygienic explications one always loses some of the features that advocates of mathematical or absolute necessity want: e.g., the hygienic notions can't be used to rule out or trivialize epistemological worries of the sort that motivate anti-realism, in the way that Lewis uses the unhygienic notion.

[23] C_T follows in a strict logical sense from $\Box AX_T$ ('$\Box$' being logical necessity). But since it is obscure what '$\Box_m$' means, how can I say what a claim that crucially involves it entails? Good question. What I meant, in saying that C_T 'more or less follows' from $\Box_m AX_T$, was simply that a discovery that C_T was false for a particular T would be regarded by any advocate of mathematical necessity as a decisive reason for denying that $\Box_m AX_T$.

They do, to be sure, involve a notion of logical possibility, and some mathematical anti-realists will not be happy with my having invoked such a notion. It seems to me however that there is no serious prospect of doing without it in one's philosophy: surely there is a difference between on the one hand characteristics (such as being red-and-not-red) which nothing logically *could* have, and on the other hand characteristics which nothing happens to have. Doubtless the epistemology of assertions of logical impossibility and logical possibility needs developing, and there are serious problems to be overcome in doing so; but I think that they are clearly *different* epistemological problems than the epistemological problems that are most characteristic of mathematics and that motivate anti-realist positions. The situation isn't entirely straightforward, because the epistemology of logical possibility is *one* element of the array of epistemological problems that arise about mathematics. For instance, *one* epistemological question about standard set theory is what reason do we have to think it logically consistent, i.e., to think that $\Diamond AX_S$? But this is not the *main* epistemological question that arises about set theory on the platonist view: there are also questions as to how we know *which* of the presumably consistent but conflicting set theories are true, and more generally, how we can know anything at all about entities in a platonic realm outside of space-time and causally isolated from ourselves and everything we can perceive. And *these* questions, which are not about logical possibility or impossibility, do not arise on the anti-realist perspective even if that perspective is taken to encompass logical possibility.[24]

6 Modality is not a General Surrogate for Ontology. I

By a natural extension of the argument in the last half of the preceding section, one could conclude that *for purposes of metalogic* there is no need for mathematical entities such as models: instead, we need only a notion of logical possibility. (Arguing this would involve arguing that the conclusions of model-theoretic reasoning, and of proof-theoretic reasoning too, can be obtained modally. See essay 3, section 4.) This

---

[24] I have heard it said that in mathematics, existence is just possible existence and truth just possible truth. Presumably what is in question in such remarks is not simply logical possibility, i.e., consistency, but mathematical possibility. Since for purely mathematical assertions it is supposed that all truths are mathematically necessary (and since the denial of a purely mathematical assertion counts as a purely mathematical assertion), the denial of a mathematical assertion requires the denial of its mathematical possibility. For logical possibility of course this does not hold.

raises the question of whether we might not avoid the introduction of mathematical entities *in science* by invoking a possibility operator.

That is, earlier on in this essay I alluded to *Science without Numbers*, in which I argued for the idea that physical theory could be developed *without any reference to mathematical entities*. The way that I proposed to develop physical theories without mathematical entities did not use modal operators either. The question then arises whether I mightn't have achieved a mathematics-free physics more easily – and whether I mightn't have had fewer difficulties in extending the programme of developing a mathematics-free physics to more complicated physical theories than the ones I considered – if I had allowed the use of modality in the development of physics.

The idea of invoking modality in dealing with physics does not strike me as an altogether appealing one. But appealingness aside, it seems to me that there is a serious problem in figuring out how an invocation of possibility could be of any *help* in formulating physical theories without mathematical entities.

It might initially be thought obvious that a notion of possibility could be invoked to avoid the postulation of mathematical entities in physics, and indeed it might be thought that it is totally trivial to turn a platonistic physics into a nominalistic physics if an operator of possibility is regarded as nominalistically acceptable. For it might be thought that if M is the mathematical theory that we use in developing our platonistic physics, then $\Diamond AX_M$ could just replace M in the application of the mathematics within science. But this is incorrect: it fails to take into account the fact that insofar as mathematics is needed at all in science, what is needed isn't just *pure* mathematical statements, but *mixed* statements which speak of mathematical entities and physical entities in the same breath. It is these mixed statements that a mathematical anti-realist must find a means to handle.

But they can't be handled by prefixing them with the modal operator '$\Diamond$': for though prefixing them with '$\Diamond$' would have the desirable effect of replacing the requirement that there *actually are* entities satisfying the mathematical theory by the weaker requirement that there *might be* such entities, it would equally have the undesirable effect of replacing the requirement that the physical world *actually* be such as to satisfy the scientific law by the weaker requirement that the physical world *might* be such as to satisfy the law. This would of course totally remove all the physical content of the law. The problem then is that it is not at all obvious that there is any way to 'modalize away' the mathematical content of the physical law (i.e., the commitment to mathematical entities) without at the same time 'modalizing away' the physical content.

There would be a way to modalize away the mathematical content of a mixed statement (such as a platonistically stated physical law) without modalizing away the physical content if we could show that the mixed statement can be represented as a conjunction of a purely mathematical component M and a purely physical component N (purely physical in that it makes no reference to mathematical entities). If such a separation could be achieved, it would allow us to reformulate the mixed statement modally as N & $\Diamond$M. It isn't hard to see, though, that the programme of separating the mixed statements of science into purely mathematical and purely physical conjuncts is precisely the programme of showing that mathematics is dispensable in science. (I announced this fact in section 2 without proof, in connection with David Lewis' attempt to undermine the epistemological objection to mathematical realism. Here, belatedly, is the proof.) For if M is purely mathematical and N is purely non-mathematical, the conservativeness of mathematics implies that the conjunction of M and N has no more purely physical consequences than N has alone; this means that if we could separate a physical law or theory into a purely mathematical conjunct and a purely physical conjunct, the purely mathematical conjunct would be dispensable without any loss of physical content to the theory. If modalizing away the mathematical content of mixed statements without modalizing away the physical content requires this sort of separation, then the introduction of modal operators, necessary though it is for dealing with metalogic, can serve no role in answering the argument that we ought to believe mathematical theories because they are indispensable to science.

It might clarify these points (and facilitate the exposition of some related ones) to look at an example. Consider a fairly typical application of mathematics to the physical world, say Laplace's equation:

$$(*) \quad \frac{\partial^2 \psi}{\partial x^2} + \frac{\partial^2 \psi}{\partial y^2} + \frac{\partial^2 \psi}{\partial z^2} = 0$$

which represents the behaviour of the gravitational field in free space in Newtonian mechanics. (Here ψ is a functor representing gravitational potential, and x, y and z are functors representing spatial co-ordinates). This obviously contains a reference to mathematical entities, so it cannot be believed by a mathematical anti-realist. The anti-realist can of course believe

$$\Diamond \left(\frac{\partial^2 \psi}{\partial x^2} + \frac{\partial^2 \psi}{\partial y^2} + \frac{\partial^2 \psi}{\partial z^2} = 0 \right),$$

but since he or she can equally believe in the logical possibility of lots of things that conflict with Laplace's equation, this is no help in providing anything that is to have the physical content of Laplace's equation without the mathematical content. (It won't do to suggest the alternative style of modalization:

$$\square\left(\mathrm{M} \supset \frac{\partial^2 \psi}{\partial x^2} + \frac{\partial^2 \psi}{\partial y^2} + \frac{\partial^2 \psi}{\partial z^2} = 0\right)$$

where M is mathematics; for of course that will be false even if Laplace's equation is correct, since even given mathematics the equation doesn't hold by logical necessity.)

Now, as it happens, there is a way to solve this problem: we can find a way to separate Laplace's equation into a purely physical part and a purely mathematical part, though it is by no means trivial to do so. The purely physical part will not involve numerical functors like the gravitational potential functor or the three co-ordinate functors: instead it will involve various comparative predicates (such as 'x is spatially closer to y than z is to w' and 'x is closer in gravitational potential to y than z is to w'); and the physical content of Laplace's equation will be stated entirely in terms of these comparative predicates without the use of set theory or any other part of mathematics. (Call this statement of the purely physical content (∗∗)). The purely mathematical part of Laplace's equation will be some part of standard mathematics – say the theory of functions of several real variables, plus a bit of set theory to guarantee the existence of sufficiently many functions from the physical world into the real numbers. If we take the statement (∗∗) that gives the purely physical part of Laplace's equation, and append it to the part of mathematics that gives the purely mathematical part, then we can derive the original platonistic version (∗).[25] This fact, together with the fact that (∗∗) is implied by (∗) taken in conjunction with some obvious 'definitions' of the comparative predicates in terms of the functors in (∗), is the precise rendering of the claim that (∗∗) states the physical content of (∗).

Now that we have represented (∗) as (∗∗) plus a mathematical theory M, the way is open for a modalist to replace M by ◇M. But one may question whether anything is gained from so doing. Note first that while (∗) follows from (∗∗) plus M, it does *not* follow from (∗∗) plus ◇M. So the modalization of mathematics does not allow us to preserve

---

[25] Or more accurately, we can derive the claim that (∗) holds for certain special co-ordinate systems and certain special methods of assigning numerical values to represent gravitational potential. (The mathematics-free version of physics makes no special reference to co-ordinate systems or to a particular choice of scale for gravitational potential, distance etc. Given the arbitrariness of any such choice, I take this to be one of the attractive features of mathematics-free physics.)

platonistic physics as literally true: if we want to preserve it as literally true, we need M itself, and hence actual mathematical objects. This leads us to the second point, which is that the real value of the decomposition is not that it allows us to modalize mathematics, but that it allows us to undercut the indispensability argument for *any* mathematics, real or modalized: (∗) is no longer something that we need believe, for we have found a way to represent its physical content directly without mathematics. Even though we don't believe it, we can use it freely in practice, for the simple reason that it follows from something that we do believe (viz., (∗∗)) plus something else that has the modal conservativeness property. So modality does come into the account of how mathematics serves to facilitate inferences in science, even if it does not come into the axiomatization of science itself. But (my third point) the modal claim that enters into *this* account is not just $\Diamond$M, it is the more complicated modal claim stating the conservativeness of M. My fourth point is a strengthening of my first: even if you took modalized mathematics to include the modal conservativeness of M rather than merely $\Diamond$M, you still couldn't derive platonistic physics (e.g. (∗)) from nominalistic physics (e.g. (∗∗)) together with modalized mathematics; you'd still need unmodalized mathematics in the derivation, so modalizing mathematics would still be of no help in licensing a literal belief in platonistic physics. And my fifth point is that the possibility of introducing a version of physics which uses no mathematics *other than* modalized mathematics depends on the fact that we have found a way to state the physical content of (∗) that relies *neither* on mathematics nor on modality. If the latter could not be done, then the introduction of modality (even modal conservativeness) could be of no help to the person who wants to do without mathematical entities.

7 Modality is not a General Surrogate for Ontology. II

The discussion above makes it hard to see how the introduction of a logical possibility operator by itself (or a mathematical or metaphysical possibility operator by itself, on any natural understanding of what that might involve) could ever be of any help in showing how to reformulate empirical applications of mathematics in such a way as to avoid commitment to mathematical entities. But perhaps we could do better if we used some other sort of possibility operator, or used a logical possibility operator in conjunction with other special logical apparatus (such as an 'actuality operator' or sophisticated variants of it). I will defer consideration of the fancy operators, and begin by discussing a modified conception of possibility in order to see if it can be used to avoid the problem discussed in the previous section.

I think that the concept of possibility which gives the idea of modalizing away the existential commitments of mathematics the best chance of working is the concept 'is a possible extension of the actual world'. Whereas the naive model-theoretic semantics for the logical possibility operator is

> $\Diamond$A is true in model M (relative to assignment function s) if and only if there is some model M' (and some assignment functions s') such that A is true in M' (relative to s'),

the naive semantics for the 'possible extension of' operator '$\Diamond_e$' in a language whose only non-logical terms are predicates is this:

> $\Diamond_e$A is true in model M (relative to assignment function s) if and only if there is some model M' which is a model-theoretic extension of M, such that A is true in M' (relative to s).[26]

(What I am calling the naive model-theoretic semantics is a semantics roughly after the fashion of Carnap 1956;[27] a Kripke-style semantics in terms of model structures containing more than one possible world could have been used instead, with no effect on the philosophical points to be made.)

At least, this naive semantics is correct for uses of the $\Diamond_e$ operator that apply to formulas involving only normal predicates; however, we will also need to introduce a rather abnormal predicate 'Act', which is to apply to all the actual objects but only to them, even when it is inside the scope of a modal operator. This means that the predicate 'Act' behaves trivially outside modal contexts: $\forall x \mathrm{Act}(x)$ is, in a sense, logically true. But it also means that $\forall x \mathrm{Act}(x)$ does not behave like a normal logical truth within modal contexts, in that for instance $\Box_e \forall x \mathrm{Act}(x)$ won't be true. In order to handle such a predicate, we need to alter the manner in which the naive model-theoretic semantics is to be specified. (The type of alteration needed is essentially that used in standard treatments of actuality operators, such as Hodes 1984a.) That is, instead of defining truth in a model, we define truth in one model

---

[26] If we use the operator '$\Diamond_e$' in a language containing individual constants and function symbols, we need to allow that closed singular terms can lack denotation in a model, and we need to understand the notion of an extension of a model in such a way that an extension of M can supply a denotation to terms that are denotationless in M. To avoid having to go through some boring details, I will for the most part stick to languages whose only non-logical terms are predicates in this section.

[27] See also the appendix to essay 3.

relative to another model that is a sub-model of the first model. I defer details to a footnote.[28] We can then define 'A is true$_s$ in M' (unrelativized to a sub-model N) as meaning that A is true$_s$ in M *relative to itself*; and we can use this to define logical truth, logical consequence, etc., in the usual way.

How might someone propose to use this apparatus in connection with the applications of mathematics? Let S be a consistent theory of mathematical physics. I will assume that S itself does not contain modal operators or the special actuality predicate; and in accordance with footnote 26 I will also make the (ultimately inessential) assumption that its individual constants and function symbols have been eliminated in favour of predicates in the usual way.[29] I will also assume that S contains a primitive predicate 'Math', meaning intuitively 'is a mathematical entity', and that it includes some assertions about which of the entities it postulates are mathematical and which are not. (If S as originally given does not include these things, we can expand it so that it does.) Now, how can we reformulate S so as not to entail the existence of mathematical entities? It might be proposed that we do so as follows:

$$(S^*) \quad \Diamond_e \, [S \ \& \ \forall x (\text{if not Math}(x), \text{then Act}(x))].$$

(Intuitively, S* says that S would hold in at least one possible world that is just like ours except in containing some extra entities *all of which are mathematical*). This avoids commitment to mathematical entities, since all quantifiers over mathematical entities are within the scope of a possibility operator and since there is no use of the actuality predicate to 'undo the possibility'. But it is not hard to see that any non-modal

---

[28] Let M be a model, N a sub-model of M and s an assignment function appropriate to M, i.e., assigning members of the domain of M to variables. Then:

1. If p is a normal predicate, $p(v_1, \ldots, v_n)$ is true$_s$ in M relative to N if and only if $\langle s(v_1), \ldots, s(v_n) \rangle$ is in the extension of p in M;
2. $\text{Act}(v)$ is true$_s$ in M relative to N if and only if $s(v)$ is in the domain of N;
3–4. Truth-functional operators as expected;
5. $\forall v A$ is true$_s$ in M relative to N if and only if for all s' appropriate to M that are just like s except in what they assign to v, A is true$_{s'}$ in M relative to N;
6. $\Diamond_e A$ is true$_s$ in M relative to N if and only if for some extension M′ of M, A is true$_s$ in M′ relative to N.

[29] Everything that isn't strictly logical must of course be made explicit prior to the elimination: for instance, if prior to the elimination of function symbols from S, S includes Laplace's equation, then before eliminating the numerical function symbols (gravitational potential, spatial co-ordinates) that appear in that equation, one must add in the background mathematics required in that equation and the background theory about the numerical function symbols.

consequence of S in which all variables are restricted to physical objects
is a consequence of S*.[30]

Does this show that we now have a way to 'modalize away' the
commitment to mathematical entities in science? There are two reasons
to doubt this. The first (which I'll call the technical reason) is that it
turns out to be possible to use this result to modalize away the

---

[30] Let an O-formula be a formula that doesn't contain the modal operator or the
special actuality predicate, and in which each variable v is restricted by the condition
'not Math(v)'; and let an O-sentence be an O-formula with no free variables. So we
need:

> Theorem 1: S* does not imply $\exists x$Math(x)
>
> Theorem 2: If A is an O-sentence and Γ is a set of O-sentences, then S* licenses
> the inference from Γ to A if and only if S does.

(To say that S licenses the inference from Γ to A is to say that A follows from S and Γ
together.) To prove these theorems we use the following obvious lemmas:

1. If M is a model of S and N$\subseteq$M and N contains all the things that fail to satisfy
   'Math()' in M, then N is a model of S*.

2. If N is a model of S* then there is an M such that N$\subseteq$M, M is a model of S, and N
   contains all the things that fail to satisfy 'Math()' in M.

3. If N$\subseteq$M and N contains all the things that fail to satisfy 'Math()' in M, and s is an
   assignment function for N and A is an O-formula, then A is true$_s$ in N if and only
   if it is true$_s$ in M.

From lemma 1 we prove that S* does not imply $\exists x$Math(x) as follows:

> Let M be any model of S. Since S entails $\exists x$-Math(x), the set of things
> satisfying '-Math(x)' in M is non-empty, so we can let N be the substructure
> of M with this set as domain; by lemma 1, N is a model of S* and of
> '-$\exists x$Math(x)', so '$\exists x$Math(x)' can't follow from S*.

From the three lemmas together we prove theorem 2 as follows:

> Suppose that A is an O-sentence and that Γ is a set of O-sentences. Then
> (a) if A is not a consequence of S* and Γ, it is not a consequence of S and
> Γ. For if N is any model of S* and Γ in which A is not true, then by
> lemma 2 there is a model M of S for which N$\subseteq$M and N contains all the
> things that fail to satisfy 'Math()' in M; and the last two conditions imply
> by lemma 3 that all members of Γ are true in M but that A is not, so that
> S and Γ fail to imply A. Conversely, (b) if A is not a consequence of S and
> Γ, then it is not a consequence of S* and Γ. For if M is a model of S and
> Γ in which A is not true, then by lemma 1 the model N defined as in the
> proof of theorem 1 is a model of S* meeting the conditions that are
> antecedents in lemma 3. So by lemma 3, all members of Γ are true in N
> and A is not, so A doesn't follow from S*.

It is worth remarking that these theorems can be carried over to languages that contain
individual constants and function symbols, providing that we introduce the complications
alluded to in footnote 26 and that we suitably redefine the notion of an O-formula.
Roughly, an O-formula must not only have all its quantifiers restricted to non-
mathematical objects, it must also contain no individual constants for mathematical objects
and no function symbols that have mathematical objects in their range.

commitment to mathematical entities in science *only on the assumption that an important part of the programme of the non-modal nominalist can be carried out.* And the second reason (the 'philosophical reason') is that the method of eliminating commitment to mathematical entities is suspiciously powerful: an analogous method would enable one to eliminate commitment to nearly all unobservables from scientific theories. I will take these two points up in reverse order.

The point to be argued first, then, is that the above method of eliminating mathematical entities is, in a sense, 'too easy': it licenses the easy elimination of far too much. (As we will see, the elimination of mathematical entities by this method is actually rather problematic – that is the substance of the technical objection that I am now deferring – but the method *does* license the easy elimination of commitment to subatomic particles, to viruses and to most other unobservables.) Suppose that T is a consistent theory (formulated without modality or the special actuality predicate) that postulates subatomic particles, and that we want to find a theory that avoids commitment to such particles but meets all uncontroversial conditions of adequacy just as well as does T. In particular, we will want it to be as observationally adequate as T. As before, we can suppose that 'subatomic particle' is a primitive predicate of T, and that T contains no individual constants or function symbols. Then in complete analogy to S*, we can formulate a modal theory T* as follows:

$$\Diamond_c[T \ \& \ \forall x(\text{if } x \text{ is not a subatomic particle then } \text{Act}(x))].$$

Then we have (theorem 1) that T* does not imply 'there are subatomic particles'; but (theorem 2) that it licenses precisely the same inferences among O-sentences that T licenses, where an O-sentence is a non-modal sentence without the special actuality predicate, in which all variables are restricted via the formula 'is not a subatomic particle'. The proof of these theorems are identical to the proof (footnote 30) of the corresponding claims for S*.

It is probably obvious how this result could be applied by a modalist to eliminate subatomic particles in a way that is *technically* unproblematic (even if philosophically quite dubious). But it is worth spelling this out explicitly: this will help in connection with my later claim that the analogous modal elimination of mathematical entities is technically (as well as philosophically) problematic. What we need to argue, then, is that T* not only avoids commitment to subatomic particles (which seems incontestable in light of theorem 1) but also that it meets all uncontroversial conditions of adequacy just as well as does T. In particular, we will need to argue that it is as observationally adequate as T. The argument for this is clear enough: presumably, anything that could be regarded as an 'observation report' for T will be equivalent (modulo T) to a sentence in which quantifiers are restricted to exclude such particles (i.e., to an O-sentence). If so, then theorem 2 gives us that T* is as observationally adequate as is T: any inference among observation reports licensed by T can be put as an inference among O-sentences, and when it is so put it is licensed by T*; conversely, any inference among observation reports expressed as O-sentences that is licensed by T* is licensed by T.

As I've remarked, I am not claiming that this vindicates anti-realism about subatomic particles; I think it reasonable to believe T. Indeed, I suspect that there can be no reason to believe T* that isn't *based on* reason to believe T: so that there is really no significant epistemological gap between T* and T. (One way to try to defend this – not the only possible way – would be to try to argue that T* is not explanatory in the way that T is; if this is right, then since the only reason to believe either theory is presumably that it explains our observations, then it is only T that could be directly supported, and T* would derive its believability from T. I think that this is a plausible line to take, though arguing that T* isn't explanatory in the way that T is is somewhat tricky.) In any case, I take it that for one reason or another most philosophers would agree that the replacement of T by T* isn't of much philosophical value. If this is so, it is hard to see why the completely analogous replacement in the mathematical case should be assumed to have any philosophical value. (For instance, it's hard to see why, if T* is less explanatory than T, S* shouldn't be less explanatory than S.)

I now turn to the technical objection: I claim that the cheap anti-realism about subatomic particles, which is at least technically possible, cannot be carried over to mathematics, at least not so cheaply. Of course, the analogues of theorems 1 and 2 still hold when T is replaced by S and the predicate 'is a subatomic particle' is replaced by 'is a mathematical entity' in defining the starring operation. But the argument for elimination in the case of subatomic particles did not consist solely in the appeal to these theorems; rather, it appealed to these theorems together with a supplementary argument that in the case of subatomic particle theory and the predicate 'subatomic particle', the class of O-sentences in the language of T is comprehensive enough to enable us to formulate our observations. (More generally, the requirement is that in going from T to the O-sentences that follow from T, one can't lose the ability to formulate what even the would-be denier of subatomic particles would want to recognize as facts.) But when we shift to a physical theory committed to mathematical entities and the use of the predicate 'is a mathematical entity' in defining the starring operation, the class of O-sentences (sentences in which all quantifiers are explicitly restricted by the formula 'is not a mathematical entity', and where if we allow singular terms in the language, singular terms for mathematical entities do not appear)[31] will not normally be comprehensive enough to formulate all of the observational facts (let alone all of the facts that the nominalist would want to recognize).

To illustrate this, suppose that S is a theory of gravitation, expressed in the usual platonist fashion. In the language of S one might express one's observations in the following way: 'the distance in meters between b and c is between 13.7 and 13.8'; 'the mass in kilograms of b is less than 17.4'; etc. These are not O-sentences: reference to real numbers is involved in their formulation.[32] I maintain that in the language typically used to formulate physics, this use of sentences that are not O-sentences to express one's observations is essential: the language does not contain O-sentences capable of doing the job.[33] If this is right, then theorems 1 and 2 are not by themselves of much use in the mathematical case: when applied to typical theories of mathematical physics, they yield theories S* that don't enable us to make the inferences about distances and masses and so forth that platonist mathematical physics allows us to make.

There is of course a way around this problem: apply theorems 1 and 2 not to a typical platonist formulation of physics, but to a modified formulation S_{mod} which contains some new predicates, chosen so that all the claims about distance, mass, velocity, acceleration etc., that we want to preserve are expressible in terms of O-sentences. The O-sentences in question *prima facie* require a very large number of infinite families of primitive predicates; a very small sample of such families might include:

(i)   the distance between x and y is r times the distance between z and w;

(ii)   the velocity of x with respect to y is r times the velocity of z with respect to w;

(iii)   the velocity of x with respect to y multiplied by the temporal difference between z and w is r times the spatial distance between u and v

---

[31] See the last paragraph of fn. 30.

[32] And if individual constants for numbers and numerical functors were eliminated in the usual way from these statements, the results still wouldn't be O-sentences, since quantification over numbers would be required.

[33] It isn't really required for my point that we regard claims about distance and mass as strictly 'observational': what is important is that they are claims that almost anyone, however nominalistically inclined, will want to find a way to preserve.

(where 'r' is to be replaced by a specific rational number; which rational number you plug in is what distinguishes one member of each family from the next.) But I will assume that our modal nominalist would not be content to employ a large number of infinite families of primitive predicates; rather, the modal nominalist will want to find a fairly small finite stock of predicates to employ as the primitives of S_{mod}, and will want to express his or her observations (and other claims he or she regards as factual) in terms of O-sentences involving this finite vocabulary.

But it is far from immediately obvious how this is to be done: in fact, the task of finding such a stock of primitive predicates and defining the members of the various infinite families in terms of them was the main technical task I set myself in *Science without Numbers*. To a large extent I solved the problem: I showed how *in the case of classical field theories in flat space-time*, and *assuming a realist attitude toward space-time points*, one *can* define from a small number of primitive predicates all of the claims about distance, mass, velocity, acceleration etc. (including how these quantities interrelate) which are needed in physics and that a nominalist would want to preserve. We can use this result, then, to solve what I've called 'the technical problem' for the modal starring method, in the case of flat space-time field theories: we apply the 'starring' operator not to S, but to a theory S_{mod} that employs a space-time ontology and the specially chosen set of primitive predicates used in *Science without Numbers*, and we express the needed observational claims in terms of O-sentences involving these special primitives in the manner done in that book.

It is noteworthy, though, that this solution to the technical problem relies both on the technical work and on some of the controversial philosophical assumptions necessary to complete the *non-modal* nominalist's programme (the programme of my book), leading one to wonder what the value is in introducing the special modal apparatus ('$\Diamond_c$' and 'Act'). It relies on the technical work: for in cases where the non-modal nominalist's programme has not yet been carried out, e.g., for theories in curved space-time, it is not obvious how to express in terms of O-sentences our observations (or other claims about the physical world that a nominalist would want to preserve, e.g., claims about the current distribution of mass in space). And it relies on some of the controversial philosophical assumptions: just as the non-modal nominalist has to assume a realist (substantivalist) view of space-time in order that his or her definitions of various concepts in terms of his or her primitive vocabulary be adequate, so too must the modal nominalist. (For instance, the modal nominalist must presuppose substantivalism if he or she adopts as spatial primitives betweenness and congruence, and expresses the claim 'the distance from x to y is 1.5 times the distance from z to w' by saying 'there are u_1 and u_2 and v such that u_1 is between x and y, u_2 is between u_1 and y, v is between z and w, and xu_1 congruent to u_1u_2 and to u_2y and to zv and to vw.' For this definition works only if it is assumed that there is *something* v midway between z and w, even if there is no matter there. To assume this amounts to assuming the existence of points of space-time. Of course, the modal nominalist could just take 'the distance from x to y is 1.5 times the distance from z to w' as a primitive predicate; but the point is that the sort of problem under discussion arises for the definition of virtually all quantitative predicates in non-numerical terms. Unless our modal nominalist is to take each member of a huge number of infinite families of predicates as primitive, I think that there is no alternative to adopting an ontology of space-time points.) In general, the promise that modality might help make it easier or more attractive to carry out an elimination of commitment to mathematical entities in the sciences appears to have vanished.

It might be contended that the problem I have been discussing arises from not taking the use of modality far enough: for, it might be said, there is nothing to stop us from formulating even our observations modally. For instance, why not just formulate the observation that the distance from a to b is approximately 1.5 times the distance from c to d by saying

> OBS $\Diamond_e$ [(i) there are real numbers r_1 and r_2 such that r_1 is the
> distance from a to b, r_2 is the distance from c to d,
> and r_1 is approximately the product of 1.5 and r_2;
> (ii) standard mathematics holds;
> and
> (iii)$\forall x$(if not Math(x), then Act(x)].

There are at least two very serious problems with this suggestion. The first is that theorem 2 of footnote 30 no longer applies: since our observation sentences are not O-sentences as previously defined, we have no reason to think that S* licenses the same inferences among observation sentences as does the platonist theory S. If the suggestion is to be of any interest, then, it must be accompanied by an extension of theorem 2 to a much more general class of sentences than O-sentences. (In fact, no such extension is possible, for the conclusion is false: this follows from the next paragraph.)

The second problem is that even putting the deductive connections among our observations aside, the suggested modal formulation of our observations is grossly deficient. Indeed, the displayed sentence OBS that attempted to state modally the distance relations between a, b, c and d in fact is true *whatever* the distance relations between a, b, c and d. Why? First, the real numbers postulated in the extended world in clause (i) are not asserted in clause (iii) to be actual; consequently, the numerical distance relation that holds in the extended world between r_1, a, and b and between r_2, c and d is not being asserted to hold between them in the actual world. (Indeed, if there are no numbers in the actual world, as the modal nominalist holds, then these relations *can't* hold there since one of their terms is absent.) Second, there is no clause in OBS that forces the distance function in the extended world to be connected in any specific way in that world with those primitive relations among physical objects which *are* required (by the semantics of $\diamondsuit_c$) to be the same in the actual and the extended worlds. Consequently the existence of a distance function with specific properties in the extended world imposes no constraints on the spatial relations in the actual world.

Can we fix up OBS so as to be immune to this problem? It is not easy to do so, short of introducing the means for a non-modal and non-platonistic formulation of the observation. For the most obvious way to fix up OBS would be to add inside the brackets a fourth conjunct that says enough about the relation between the numerical distance function and some primitive physical predicates so as to determine an explicit non-modal definition of distance relations like (i) in terms of these primitive predicates. We see again that if we solve the non-modal anti-platonist's problems, we solve the modal anti-platonist's too: but only by making the appeal to modality otiose.

Perhaps, though, there is another way to improve on OBS: one that uses a still more powerful modal apparatus, such as a full actuality operator in conjunction with a quantifier over relations. This would enable us to add a different kind of fourth clause to OBS, not requiring new predicates in the language:

> OBS$^+$ $\diamondsuit_c$[(i) there are real numbers r_1 and r_2 such that r_1 is the distance from a to b, r_2 is the distance from c to d, and r_1 is approximately the product of 1.5 and r_2;
> (ii) standard mathematics holds;
> (iii) $\forall x$(if not Math(x), then Act(x)).
>
> and
>
> (iv) $\forall R \forall x_1 \ldots \forall x_4$[if Act$(x_1)$ and . . . and Act(x_4) then [$(R(x_1, \ldots, x_4)$ if and only if actually R$(x_1, \ldots, x_4)$]].[34]

---

[34] The Act predicate is of coure definable in terms of the actually operator, as 'actually $\exists y(y=x)$.'

In (iv), R ranges over 4-place relations. (We should really add a clause similar to (iv) for relations of all other finite adicity, but for sake of readability I will stick to adicity 4.) In effect what we're doing here is introducing a more powerful operator $\Diamond_c{}^+$, where $\Diamond_c{}^+A$ means that A holds in some enlargement of the actual world in which *all* relations among actual entities are preserved, not just those relations expressible in atomic predicates in the original language; and we are replacing $\Diamond_c$ by $\Diamond_c{}^+$ in OBS.[35]

Is OBS$^+$ any better than OBS? Basically the answer is 'no'; but to say why, one has to distinguish between different readings that could be given to the relation quantifier in the new actuality clause (iv).

1   If we understand the relation quantifier in OBS$^+$ as ranging over *all* relations, or even all relations definable in platonist physics, then OBS$^+$ is of no use whatever to the modal nominalist, for it does not avoid the commitment to mathematical objects. For clause (i) of OBS$^+$ expresses a platonist spatial relation between a, b, c and d: the relation R_i of there being numbers r_1 and r_2 such that $r_1 = \text{dist}(a,b)$ and $r_2 = \text{dist}(c,d)$ and $r_1 = 1.5 \times r_2$. If the relation quantifier in (iv) ranges over *all* relations, even platonistic ones, then part of the content of (iv) (given that a–d are actual, by (iii)) is that a, b, c and d stand in R_i in the actual world if and only if they stand in it in the extended world. Consequently, clause (iv) allows us to export the commitment to numbers in (i) from the extended world back to the actual world. We have succeeded in making OBS$^+$ constrain the actual spatial relations among a, b, c and d; but we have done so only at the cost of making it as committed to mathematical entities as the platonist formulation of the observation statement from which we started.

2   Clearly, then, if OBS$^+$ is to serve our purposes we must restrict the relation quantifier in (iv) to *non-platonistic* relations among physical

---

[35] We would do slightly better in defining $\Diamond_c{}^+$ (or formulating OBS$^+$) to use, not the standard actuality operator, but Hodes' 'backspace operator' (Hodes 1984a). This operator behaves just like the standard actuality operator until one comes to contexts governed by *multiple* modal operators, but in those contexts it behaves in a more useful fashion. (What it does, roughly, is to 'temporarily undo the effects of' the most immediately governing modal operator, whereas the standard actuality operator 'temporarily undoes the effects of' *all* the modal operators governing the context in question: as Hodes remarks, the standard operator is like a carriage return instead of a single backspace.) Use of the backspace version of 'actually' instead of the more standard carriage return version has two advantages: it makes $\Diamond_c$, and thus OBS$^+$, better behaved inside modal contexts; and it means that we needn't take $\Diamond_c$ as a primitive in formulating OBS$^+$, we can make do with the standard logical possibility operator $\Diamond$ together with the backspace version of 'actually', since $\Diamond_c$ is definable in terms of the backspace version of 'actually' though not the carriage return version. These are matters of detail that will not affect the points to follow.

objects – i.e., to relations among physical objects that 'do not in any way involve' mathematical objects. A natural way to interpret the restriction to 'relations that don't in any way involve' mathematical objects is as a restriction to relations definable in the vocabulary of physical theory without reference to or quantification over mathematical entities. But it is easy to see that if this is how we interpret it, then OBS^+ becomes equivalent to OBS. In other words, on this construal of the relation quantifier, OBS^+ is not committed to mathematical entities, which is good; but it also does nothing to constrain the spatial relations among objects. It seems as if one cannot properly constrain the spatial relations among objects without introducing mathematical entities – at least, we can't do this unless we can solve the problem of the *non-modal* anti-platonist.

In essay 6, in discussing the prospects of using a modal operator like $\diamondsuit_c^+$ to eliminate ontological commitment to space-time, I made a distinction between predicative and impredicative treatments of the quantifier over physical relations, and conceded that on an impredicative construal the modal translation could not be shown to be extensionally incorrect. (Though, I argued, it had other devastating problems.) It is worth noting that the case of OBS^+ is different: here, making the relation quantifier impredicative wouldn't even solve the problem of demonstrable extensional inadequacy. For OBS^+ still contains nothing that forces the distance function in the extended world to be connected in any specific way in that world with the non-platonistic relations of that world (that is, with the relations, whether definable in the language or not, whose holding between two objects does not entail the existence of mathematical entities). Consequently, even if we treat the quantifier as ranging over impredicative relations, we have a dilemma: either those relations include platonistic relations, in which case OBS^+ collapses into the platonistic account of distance relations; or those relations don't include platonistic relations, in which case OBS^+ provides no constraint at all on the spatial relations among a, b, c and d.

The disanalogy with essay 6 suggests that there might be a slightly better modal definition than OBS^+, one that gives the impredicativity move a useful role to play. Perhaps so, but I see little reason to pursue this. After all, the problem we have been pursuing in the last few paragraphs – the problem of OBS^+ coming out true when it shouldn't – was only one of the two serious technical problems that arose for the idea of construing our observations modally. The other problem, that there is no reason to expect theorem 2 to extend to this enlarged class of observation sentences, would still arise in full force.

It should also be noted that the use of the more powerful modal operator $\diamondsuit_c^+$ instead of $\diamondsuit_c$ could be of no use independently of construing our observations modally. That is, there is no hope of solving our earlier problems by going back to the idea of using O-sentences that expressed observations non-modally (so that we get theorem 2), and just using $\Diamond_c{}^+$ instead of $\Diamond_c$ in the formulation of our modal physical theory S*. For the technical problem with S* was that it did not enable us to express our observations as O-sentences, and the introduction of the more powerful modality can obviously do nothing to solve that.

It is worth reiterating that even if the modal nominalist could find a way around the sort of technical difficulty I have been discussing (and even if his or her way around it did not simply ride piggyback on the work of the *non*-modal nominalist), it is not clear that much would have been gained. All that such a technical success would show is that there is a way to avoid the special technical difficulties that seemed to beset the modal elimination of mathematical entities and that don't arise for the modal elimination of subatomic particles and viruses. In other words, the solution to the technical problem would only show that we can have a modal elimination of mathematical entities that is every bit as good as the modal elimination of subatomic particles and viruses. Almost everyone would agree, I think, that that conclusion should be of little solace to the nominalist. If modality is to be useful in eliminating the commitment to mathematical entities in science, it must be employed in some less trivial way. I doubt that there is a way, both workable and non-trivial, to modalize away the mathematical content of empirical science without at the same time modalizing away its empirical content (unless of course one relies on a non-modal nominalization); in any case, if there is such a way, it would have to be very different from any I have considered here.

8 Putnam on Realism, Mathematics and Modality

A

What I have been saying in the previous two sections creates a difficulty – perhaps not an insurmountable one – for the position that Hilary Putnam has taken in various papers such as 'Mathematics without foundations' (1967b) and 'What is mathematical truth?' (1975). In these papers Putnam argued that there was no need for a mathematical realist to adopt the 'mathematical object picture' – the realist could formulate his or her views in purely modal terms – and that this wasn't an *alternative* to the mathematical object picture but was the same view put in a different way. Putnam also argued (at least in the later paper) that the primary argument for mathematical realism was the argument that mathematics is indispensable to physical science. He illustrated the apparent indispensability of mathematics to science by the sort of equation I've mentioned (Laplace's). Now, it seems indisputable that such equations provide a *prima facie* argument for the indispensability of *mathematical objects* to science (though I think that the restatement of the equation in terms of comparative predicates ultimately undercuts the argument). Consequently, if 'mathematics as modal logic' really were an equivalent description of mathematics as presented in terms of mathematical objects, it should be possible to reformulate the indispensability argument so that it is a *prima facie* argument for the need of modalized mathematics (or, a *prima facie* argument for the need of *one or the other* of modalized mathematics and mathematical objects). But the moral of sections 6 and 7 would seem to be that we cannot so formulate the indispensability argument: rather, insofar as the indispensability argument works at all, it requires mathematical objects, and modalized mathematics will not fit the bill. We have to be a bit careful in drawing this conclusion: I did not attempt in sections 6 and 7 to consider every possible modal version of the application of mathematics, and Putnam does not commit himself to the particular versions I did consider. I do not think it likely, though, that Putnam will find a way around the problems I have raised. If not, modalized mathematics cannot be an equivalent description to mathematics as normally presented. (This fact partly explains why I regard the acceptance of a logical possibility operator (for purposes of metalogic but not for purposes of reformulating science) as not undermining my claims to be advancing a mathematical *anti*-realist position.)

B

Even independently of worries about the problem of application, Putnam's claim that his 'mathematics as modal logic' is a form of mathematical realism may seem puzzling. To discuss this I will need to say a bit more about what his 'mathematics as modal logic' consists of. Putnam is claiming that there is some sort of modal translation of pure mathematics: he envisions finding a translation procedure that takes mathematical assertions into modal assertions, one that takes *acceptable* mathematical assertions (e.g., the axioms of set theory) that quantify over mathematical entities into *true* modal assertions that involve no such quantification unless it is modalized away. (In other words, the modal assertions cannot straightforwardly imply the existence of mathematical entities.) Presumably Putnam thinks of the translation procedure as applying to impure mathematical assertions as well as to pure ones. The earlier discussion of the problem of application indicates that there is a serious obstacle to getting a translation procedure that applies to the impure mathematical assertions of a science for which we do not already have a non-modal nominalization. But I shall have no more to say about this, so let us focus our attention on the pure mathematical assertions only. Even for such pure mathematical assertions, Putnam is not entirely clear about exactly how the translation procedure is to go, but I do not want to discuss that here. But there are two general questions about the translation scheme that I do want to discuss briefly, since both relate to the issue of realism: the first question concerns the nature of the modality that is involved in the translations, and the second is the question of what *value* such translations have.

On the first point: Putnam regards 'mathematics as modal logic' as in some sense an alternative to 'the mathematical object picture', but thinks that, at a deeper level, the two are equivalent descriptions. It seems to me that there is an interesting and an uninteresting way to regard what he says, but that unfortunately the uninteresting one is almost forced on us if we take seriously a remark he makes in 'What is mathematical truth?': the remark that mathematics as modal logic employs 'a strong and uniquely mathematical sense of "possible" and "impossible"' (p. 70).[36] Why does this remark almost force an uninteresting interpretation? Note first that on any reasonable understanding of the phrase 'mathematically possible', mathematical possibility coincides with purely logical possibility *as applied to purely non-mathematical statements*: that is what the conservativeness of mathematics is all about.[37] Consequently, if there is any point to insisting that the modal translation of a mathematical assertion employs 'a strong and uniquely mathematical notion of possibility', that translation must attach a mathematical possibility operator to some statement A that entails the existence of mathematical entities. It is plausible to suppose that this possibility operator would be the main connective of the modal translation; if so, we have that the modal translation has the form $\Diamond_m A$ *where A entails the existence of mathematical entities*. A might be *impurely* mathematical; but if so, since it involves mathematical entities it will entail a purely mathematical statement B, and so our modal translation $\Diamond_m A$ will entail $\Diamond_m B$ where B is purely mathematical.

[36] Putnam (1967b) said that only a purely logical modality was employed. I do not know why he shifted away from that position.

[37] More fully: if A is not mathematically possible, then '-A' is a consequence of mathematics; so if A (and hence its negation) are purely non-mathematical, then '-A' is logically true.

But for *purely* mathematical statements, mathematical possibility and truth coincide: cf. footnote 24. Consequently, it would seem, the insistence on using 'a strong and uniquely mathematical notion of possibility' means that the modal translations of mathematical assertions are every bit as much ontologically committed to mathematical entities as are the mathematical assertions themselves, and there is no serious sense in which mathematics as modal logic employs an alternative to the mathematical object picture.

Given this, it seems best to try to disregard the remarks in Putnam 1975 about 'mathematics as modal logic' employing 'a strong and uniquely mathematical notion of possibility'.[38] If we do so, there is no difficulty in understanding how, if a modal translation of pure mathematics is possible, it offers a genuine alternative to the 'mathematical object' picture of mathematics. (Nor, as we shall see in footnote 40, is there any difficulty in seeing how there is a sense in which – that is, a construal of the mathematical object picture in which – the two pictures might be equivalent descriptions). This, then, is how I shall try to interpret Putnam in what follows.

C

My second question (or class of questions) about Putnam's modal translations is, what is the point of giving them? What is the value of a translation procedure that associates true modal statements with acceptable statements of mathematics? Does such a translation scheme somehow confer truth (as opposed to mere acceptability) on the mathematical statements themselves? If so, what is the value of saying that the mathematical statements are true as well as acceptable? And so forth.

In order to approach these questions, it is helpful to look at Putnam's remarks on mathematical realism. Putnam regards himself as a 'mathematical realist'. Earlier in this essay I explained what *I* meant by 'mathematical realism', by saying that mathematical realism is the doctrine that there are mathematical entities that are mind-independent and language-independent. Putnam (1975) however uses the term 'realism' differently; he says:

it is possible to be a realist with respect to mathematical discourse without committing oneself to the existence of 'mathematical objects'. The question of

---

[38] Perhaps Putnam meant to say, not that the notion of possibility involved was to be mathematical possibility as opposed to logical possibility, but that it was to be *second order* logical possibility as opposed to first order? On that interpretation, the objection of the previous paragraph would not apply.

realism, as Kreisel long ago put it, is the question of the objectivity of mathematics and not the question of the existence of mathematical objects. (p. 70)

This remark can seem puzzling: it is easy enough to see how to interpret the Kreisel dictum as requiring *more* of the mathematical realist than the belief in mathematical entities (that is how Michael Dummett (1978, pp. xxviii–xxix) and Crispin Wright (introduction to 1983) interpret it);[39] but it is not so obvious how to interpret it so as to require any less. The *prima facie* difficulty is as follows.

First of all, I take it that on Putnam's (and Kreisel's) view 'mathematical realism' involves not merely that there is objectivity as to what mathematical conclusions follow from what mathematical axioms, but also that there is objectivity as to which mathematical axioms to accept. (If only the former were at issue, we wouldn't be concerned with the objectivity of *mathematics* but only the objectivity of *logic*; and the view would be rather similar to the 'if-thenism' that Putnam now rejects.) And the natural way to explain the objectivity as to which axioms to accept is this: some of the purported axioms are (at least disquotationally) true, the others are not; and the objectivity consists in a constraint that we accept only the true ones. But many of the mathematical axioms that we accept are existential: e.g., there is a natural number 0, and for every natural number there is a greater one. If these axioms are disquotationally true, there are natural numbers – and natural numbers are the paradigms of mathematical objects. So the objectivity of mathematics (if interpreted as requiring at least the disquotational truth of acceptable mathematical axioms) requires the existence of mathematical objects. That's the *prima facie* difficulty.

Putnam cites his 'mathematics as modal logic' as an illustration of Kreisel's dictum. As I've said, the idea is that claims involving quantifications over mathematical entities are to be translated away into modal claims of some sort that don't involve such commitments – with acceptable mathematical claims translated into true modal claims, and unacceptable mathematical claims translated into false modal claims. If this programme can be satisfactorily carried out and is taken to be of value, then there are two different (at least, verbally different) ways to regard the translations: we can regard them as genuine equivalences, in which case truth of the modal translations shows the truth of the platonistic sentences (the ones existentially quantifying over mathematical

---

[39] The added content they give it is closely related to, but not quite the same as, the added content that I gave to realism by requiring that mathematical entities be mind- and language-independent. See Dummett (1978, pp. xxv–xxix and 228–30).

entities); or we can regard the modal translates as the genuine truths, and the platonistic claims as literal falsehoods that in many contexts we can regard as truths without harm since there is no danger in those contexts of the difference between them and their literally true modal counterparts being put to use. (E.g., if we are deciding how to build a bridge, we will be led to the same decision in either case, so that the platonistic claim, even if not strictly true, can be regarded as true for the purposes at hand. But if we are deciding whether to believe in mathematical entities, the two assertions lead to different conclusions; and here, according to our second way of regarding the translation, the difference in truth value between the platonistic assertion and its modal translation becomes manifest.) In my view it makes absolutely no difference which of these views we adopt: they differ only on the verbal issue about the use of the word 'true'.

If this is right, the argument that objectivity as to the choice of mathematical axioms requires the existence of mathematical objects should either stand on both views of the translation or should fall on both views. And in fact it falls on both views, for on both views we have the required objectivity as long as it is objective which modal assertions to accept. If we accept the second view of the translations, it is easy to see where our argument that Kreisel-objectivity required the existence of mathematical entities went wrong: it went wrong in the assumption that the objectivity in the choice of mathematical axioms had to be explained in terms of their truth; if an explanation of the objectivity in terms of the truth of *associated modal assertions* were given instead, the argument would break down. If on the other hand we adopt the first view of the translations, i.e., regard them as truth-preserving, then we have to interpret 'true' non-disquotationally if we are to avoid mathematical objects: 'true' can only mean 'comes out disquotationally true on the modal translation', for otherwise the truth of existential claims in mathematics will imply the existence of mathematical objects. But if we do regard 'true' in this non-disquotational sense, we again have a way around the *prima facie* argument that Kreisel-objectivity requires mathematical objects.[40]

[40] There is still a third closely related view of the translations (or reformulation of the same view of them): that the translation scheme in question preserves disquotational truth, so that mathematical entities do exist; but that the existence of these mathematical entities is somehow derivative from the truth of the modal claims. (This last clause is necessary to distinguish the view from a more thorough platonism.) Such a view (or reformulation of the view in the text) can be used to raise doubts about whether the existence of (even mind- and language-independent) mathematical entities should be

cont. over page

Returning now to the question of what value modal translations of mathematics might have (supposing acceptable translations to be forthcoming), it should be clear that we do not get a helpful answer to this question if we simply say that the modal translation of mathematics shows that *acceptable* mathematical assertions are also *true*. For whether we regard such a translation procedure as conferring truth on the mathematical assertions or not is a purely verbal issue. Even assuming the existence of a scheme of modal translation, it is perfectly possible to describe the resulting view as a form of instrumentalism about mathematics: it can be put by saying that mathematics taken at face value is not disquotationally true, but that it needn't be disquotationally true to be good (it need only have a disquotationally true 'translation').

What then is the value of the search for modal translations (or any other sort of translations of mathematics into acceptable nominalistic terms)? Why not instead adopt the somewhat easier course of simply trying to translate each of the applications of mathematics? (This is in effect the task that my own nominalization programme is committed to.) Such a 'translation' of each application of mathematics would of course be forthcoming from an acceptable procedure for translating

---

regarded as a *sufficient* condition for realism; whereas Putnam was raising doubts about whether it should be regarded as a *necessary* condition for realism.

It seems to me that the view described here is most naturally regarded as a redescription of the views in the text. (So I partially agree with Putnam: I am saying that if we assume for the sake of argument that it is possible to find a modal 'translation' of all mathematical assertions, pure and applied, then we *can* view the translation as preserving disquotational truth, and hence *can* understand talk of the existence of mathematical objects in such a way that it is disquotationally true and an equivalent description to modal talk.) If it seems puzzling that two equivalent descriptions could differ in the existential claims that follow from them, the answer is that they employ the word 'exists' in different senses. Under this third redescription of the views in the text, our 'epistemological access to the mathematical realm' is as unproblematic as our modal knowledge. (Of course, there are issues as to how unproblematic that is; but if the modality at issue is a purely logical modality, then whatever problems there are are different from the usual problems about the existence of a platonic realm, as I argued at the end of section 4.) But if one insists that this third view is to be understood as not a redescription of the views in the text (as Wright (1983) insists when he maintains this view of the translations *while explicitly opposing* what he calls 'ontological reductionism'), then there remains a gap between our modal knowledge and our knowledge of mathematical entities; there is a significant epistemological question concerning our knowledge of the 'translation scheme'. So I think that the third view of the translations, coupled with the insistence that it differs substantively from the first and second, should be regarded as a form of platonism.
form of platonism.

I should add that I do not think that this third view of the translations is philosophically very useful. The reason is that even if it in some sense gives us mathematical entities that make pure mathematics come out true, it can't do the same for applications of mathematics; at least, not without independently solving the problems of section 7.

mathematical statements themselves, mixed as well as pure. But the task of trying to 'translate' each of the applications of mathematics without invoking mathematical entities is somewhat easier than the task of trying to translate the mathematics itself, since in translating each application there is no requirement that you translate the purely mathematical claims themselves (or that if you do, that you translate them the same way in connection with each application).

I think that the real answer to the question of why Putnam regards modal translations as having value can be stated independently of talk of the truth of mathematics: the answer involves the issue of the objectivity in the choice of mathematical axioms. That is, the idea is that if we can come up with an acceptable scheme for modally translating mathematical assertions, this will provide a strong constraint on which mathematical axioms to accept: we shall accept as axioms only those mathematical assertions that have disquotationally true modal translations. Recall that on the view I advocated in section 3, the only demand on a mathematical theory (aside from such requirements as that it be powerful enough and elegant enough to be interesting) is that it be conservative. But this is not a very strong constraint on the choice of a set of mathematical axioms. For in the first place, conservativeness (like consistency) is a holistic property of theories, not explainable in terms of the conservativeness of the individual axioms that comprise it. The same axiom may be acceptable in one theoretical context but not another, in that the theory consisting of that axiom plus the other assumptions of the first theoretical context is conservative but the theory consisting of that axiom plus the other assumptions of the second context is not conservative. Disquotational truth is different from this: the disquotational truth of a theory is explainable as the disquotational truth of each component axiom. In the second place, it is often possible to add either an axiom or its negation to a theory, and get a conservative theory either way: I mentioned before that if standard set theory is conservative, so are both standard set theory plus the general continuum hypothesis (GCH) and standard set theory plus its denial. Again, disquotational truth is different: from a platonist point of view there is an objective choice to be made here between GCH and its denial, namely, we should choose the true one. And Putnam's view preserves this objectivity in platonism (or so it seems at first blush anyway): on Putnam's view it is an objective matter as to whether GCH has a true modal translation. This may seem a very attractive feature of Putnam's view over mine.

The first thing I want to say about this is that the description in the previous paragraph of the difference in objectivity between my view and Putnam's is considerably over-drawn. Indeed, the description there

of the difference in objectivity between my view and the platonist's
view is over-drawn as well. For my description in the previous paragraph
slurred over one source of non-objectivity in the platonist view; and it
slurred over two sources of non-objectivity in Putnam's view, one of
them pretty much analogous to the source of non-objectivity in the
platonist's view and one additional.

The source of non-objectivity in the platonist view is the fact that
prima facie conflicting mathematical theories needn't be genuinely in
conflict: each theory could be true *on the interpretation appropriate to
the advocates of that theory*, for the advocates of one theory and the
advocates of the other could attach different truth conditions to the
same sentences. Consider two platonists, X_+ who advocates GCH and
X_- who denies it. There is no need for X_- to regard X_+ as wrong: as
Gödel's relative consistency proof shows, X_- can reinterpret the
assertions of X_+ so that they actually follow from X_-'s own; indeed,
he or she can do so simply by interpreting X_+'s quantifiers as restricted
(by a restriction specifiable in X_-'s language), leaving the interpretation
otherwise homophonic. Consequently, it is natural for X_- to say not
that X_+ is wrong but simply that X_+ has a less inclusive notion of set.
Can X_+ also admit that X_-'s assertions might be true? Yes, in various
ways. First, let's first make things difficult, by supposing that X_+ thinks
that he or she has the widest possible notion of set (so that GCH is
true on this widest possible notion). Even such a person can admit the
possibility that there is a suitable restriction of quantifiers (one that
eliminates lots of 1–1 correspondences between sets that are not
themselves eliminated) such that with this restriction of quantifiers not-
GCH holds even if the interpretation of 'ϵ' is held fixed. Indeed,
Cohen's relative consistency proof partly establishes this possibility:
given only the additional assumption that there is a standard model in
which the ZF axioms hold (a consequence of 'plausible' axioms like the
axiom of inaccessibility), there is a standard transitive model in which
ZF + not-GCH comes out true, and this can be used to provide an
interpretation of X_-'s assertions that makes them all true and that
departs from homophonicity only in involving restricted quantification.
(In this case, the restriction is specifiable in X_+'s language only relative
to a parameter, but it is hard to see why that should matter.)[41] Secondly,

---

[41] Even without the assumption of a standard model, X_+ can get such an 'almost
homophonic' interpretation of *arbitrarily large finitely axiomatized fragments* of ZF +
not-GCH; alternatively, he or she can specify a Boolean-valued interpretation (or an
interpretation in terms of forcing conditions) in which the full ZF + not-GCH comes
out true. Also, without the assumption of a standard model but assuming the formalization
of ZF's consistency in ZF, X_+ can get that there is a model of ZF + not-GCH; this
shows that there is a (normal as opposed to Boolean-valued) interpretation, though in
this case the interpretation of 'ϵ' needn't be homophonic.

further possibilities are opened up if X_+ is not committed to the view that his or her own conception of sets is the most inclusive one possible. And there is no reason why an advocate of any set theory should hold that his or her own conception of sets is the widest possible;[42] the advocate of the set theory need only hold that the set theoretic claims he or she accepts (such as perhaps GCH) are true of sets as he or she conceives them.[43]

The moral is clear: even if we assume that the interpretation of 'ϵ' is fixed, the platonist view of set theory does not demand that the truths of set theory be objectively determined: for there are different totalities that the quantifiers in a set theory could range over, and which set-theoretic claims are true will depend on which totality is chosen. Indeed, the moral can be strengthened, for there is no obvious reason to regard the interpretation of 'ϵ' as fixed. This has been well argued by Putnam himself, in 'Models and reality': Putnam argues convincingly in the early sections of that paper that the only thing that could help in fixing which interpretation of 'ϵ' is 'intended' by a person or a community is the acceptance of sentences containing 'ϵ' by that person or that community.[44]

[42] To hold anything like that would in fact seem rather implausible. Look at the way other mathematical concepts such as 'number' have admitted successive expansions: the introduction of negatives, fractions, irrationals, imaginaries, Robinson infinitesimals, transfinites, surreal numbers (Knuth 1974) etc., with no obvious reason to think that an end is in sight. Why shouldn't the same happen for sets?

[43] It might be thought that to the extent that we rely on this last point, the objectivity of mathematics on the platonist conception is granted: for there is a certain advantage in choosing as wide a conception of sets as possible, and if there is an objective question as to whether GCH holds *on the widest possible conception of set*, then that gives an objective answer to the question of the GCH on the platonist conception. There are two replies to this. First, it is not in the least obvious that there is a 'broadest possible notion of set', nor that if there is one it is unique – indeed, that seems doubtful, as I argued in fn. 42. I see no grounds to rule out the possibility that for each conception of set on which one of (GCH, not-GCH) hold, there is an at least equally broad one on which the other holds. Second, recall that the context of this discussion is the issue of whether mathematics is more objective on a platonist view than on a view according to which the requirement that mathematics be true is dropped in favour of the requirement that mathematics be conservative. But the strategy of arguing that (say) not-GCH is better than (say) GCH, not because the first is true and the second false but because the first is true on a broader conception of set, is easily adaptable to a view in which conservativeness replaces truth. See the next to last paragraph of this subsection.

[44] In later sections of the paper, Putnam extends his conclusion to non-mathematical predicates. The extension is fallacious, I believe: it rests on confusing the view that reference is determined by e.g. causal considerations with the view that reference is determined by a description theory in which descriptions containing the word 'cause' contain an especially prominent role. (See Glymour 1982; Devitt 1983; Lewis 1984.) But the dubiousness of the extended view does not undermine the view in the case of mathematical predicates, where nothing like causal considerations can play a role in determining which interpretation(s) are intended.

If only the acceptance of purely mathematical sentences is at issue, then any interpretation under which a mathematical theory is true meets all the conditions for being an intended interpretation; so that any consistent mathematical theory comes out true on an interpretation intended by its advocates. If mixed sentences are considered as well, the situation is more complicated, but I think it plausible that the same thing holds except with 'consistent' replaced by 'conservative'. If this is right then the platonist view of mathematics does not in fact make mathematics any more objective than does the view I have advocated. And even if what I've said is not right in detail – even if there are more constraints on what it is for an interpretation to be 'intended' – still the fact that different mathematicians might mean different things by their words casts doubt on the idea that the choice of mathematical axioms on the platonist picture comes out completely objective in Putnam's sense.

When we move from the platonist picture to Putnam's picture, there is room for even more doubts as to how far Putnam's ideal of objectivity would be achieved. In the first place, there is a doubt analogous to the doubt that arises in the case of the platonist position. Suppose two mathematicians disagree about the continuum hypothesis, but both are advocates of Putnam's 'mathematics as modal logic', and indeed both agree on the translation procedure for translating mathematical claims into modal claims. Consequently, both agree that their mathematical disagreement about the continuum hypothesis amounts to a modal disagreement about a certain modal claim S (the modal translation of the continuum hypothesis). But just as with the platonist construal of mathematics, there need be no fact of the matter as to which mathematician is right about S: it could well be that S in the mouth of one mathematician differs in meaning from S in the mouth of the other, so that when one affirms it and the other denies it both are speaking the truth. I think that this is more than an abstract possibility (though it is hard to establish conclusively that it arises in absence of a detailed proposal as to what the modal translation looks like). One thing that is clear about the modal translations is that they contain modal operators. For reasons I shall not discuss, it seems quite unlikely that Putnam would interpret these as representing purely first order logical possibility (i.e., purely first order logical satisfiability). Indeed, as we've seen, Putnam himself says in his most recent article on the subject that they don't represent logical possibility at all, but a special sort of mathematical possibility. I don't think he *should* say this, as I've argued, but I think that the most natural fallback position is that they represent satisfiability either in a higher order logic or in an enrichment of first order logic that includes such things as infinitary quantifiers or devices to represent the notion of a well-ordering. These modal notions do not of course admit of complete proof procedures. What reason is there to think that when one mathematician uses them he or she means exactly what another one means by them? Why shouldn't each mathematician interpret the other's modal operator nonstandardly? If the arguments of 'Models and reality' are sound, they apply here to show that such a nonstandard construal is appropriate and that as long as each mathematical theory is consistent (or conservative, in the case of impure theories) then the 'intended' interpretations of it are such as to make it true.[45]

So far the discussion of the illusoriness of the 'objectivity' of the modalist position is closely analogous to the discussion of the illusoriness of the 'objectivity' of the platonist position. But in fact there is a second source of doubt about the 'objectivity' in the choice of axioms on the modalist position, one that arises even if one pretends that there is a unique meaning of the modal operator (and of any special predicates used in the modal translations), a meaning which all who use it grasp whatever their doctrinal disagreements about modality. Recall that the 'objectivity' that arises for the modalist view arises from the constraint that we shall accept as axioms only those mathematical assertions that have disquotationally true modal translations. But even aside from the fact that I have been emphasizing, that there may be no fact of the matter as to which model translations are true, there is another fact to emphasize: that there are *many different* ways one might modally translate a given mathematical axiom. In the discussion above I assumed that the advocate of the continuum hypothesis and the advocate of its denial would agree on how the continuum hypothesis is to be modally translated. But why should we suppose that? If we don't suppose it, we have still further reason to doubt that the modalist view achieves any more 'objectivity' about the choice of mathematical axioms than does the view I outlined in section 3.

Although I am skeptical that one can achieve the sort of objectivity that Putnam wants in the choice of mathematical axioms by imposing the demand that an acceptable mathematical axiom have a disquotationally true modal translation, I do not pretend that the discussion of the past few paragraphs shows conclusively that such objectivity cannot be so obtained. Doubtless further investigation of these matters is called for. Another matter that deserves further investigation is whether Putnam's proposal for how objectivity in the choice of axioms is to be achieved is well motivated. Suppose that we want to attach an important sense to the claim that 'real mathematical possibility' corresponds to consistency

---

[45] If the modal translation contains special primitive predicates as well as a modal operator, there is even more room for maintaining the position that apparent disagreements are illusory.

with standard set theory S, rather than consistency with a nonstandard but equally conservative set theory S′. It seems clear that if there is sense to this, it must be *given*. Putnam proposes a programme for giving sense to it: find a certain sort of modal translation, in such a way that one set theory but not the other has a true modal translation. But even if Putnam's programme were carried out, would it really give sense to the claim that S has an importance that S′ doesn't? It seems to me that it would only show this if it were shown *why modal translations of the sort in question were of particular interest*. One way to try to show that would be by a theory of application: showing that a decent account of application had to rest on a general modal translation. I've tried to give an account that doesn't require that, though (and also to suggest difficulties for accounts like Putnam's that do require it); so unless my account of application is inadequate (and my doubts about modal accounts of application misplaced), it seems to me that the modalist programme for achieving objectivity in the choice of axioms lacks obvious motivation.

I should conclude this discussion of the objectivity in the choice of set-theoretic axioms by noting that my own view can give a certain sort of sense to the claim that one set theory S is 'better than' an alternative but equally conservative set theory S′. For I stressed before that just as a platonist values more in a mathematical theory than truth, I value more in a mathematical theory than conservativeness – I, like the platonist, value such things as powerfulness. Now, what sort of powerfulness do we want in a set theory? A main purpose of set theory in modern mathematics is to serve to represent possibilities: for virtually any theory that we take to be consistent we can give a model of it in set theory;[46] indeed, virtually any theory we take to be consistent can be modelled in a particularly vivid way, by embedding it in set theory (giving an inner model or inner interpretation). Given two set theories both of which are conservative, one that enables us to model more theories, and model them more vividly, is more useful for these purposes than the other. I think it is this fact that explains why mathematicians tend to prefer set theories with powerful large cardinal axioms to set theories without them, and why they prefer set theories that deny the general continuum hypothesis to set theories that assert it. There is no

---

[46] This fact depends not only on the fact that set theory is powerful enough for a completeness proof: for (a) only if the set theory is also powerful enough to establish the proof-theoretic consistency of the theory in question does the completeness theorem show the existence of a set-theoretic model; and (b) using the completeness theorem to prove that there is a model of a theory does not show that 'we can give' a model of the theory, because the completeness theorem is proved so non-constructively.

reason, though, to think that the desideratum on a set theory of enabling us to model lots of theories and do so in a vivid way picks out one set theory uniquely. Nor is there reason to think that a set theory that does not do well on this desideratum can't in many other contexts be a perfectly good mathematical theory. The comparative virtues of the set theories we tend to like over those we tend not to like can be explained without subscribing to the myth that one set theory stands out over the others as 'the true one'.

In short, I think (a) that it is highly questionable that we need any sort of objectivity in the choice of mathematical axioms that is not explainable in terms of the view offered in section 3; (b) that there is no obvious motivation for regarding having a disquotationally true modal translation as relevant to any sort of objectivity worth caring about; and (c) that there is no obvious reason to think that the requirement of having a disquotationally true modal translation successfully discriminates among mathematical axioms anyway. These doubts are additional to my main doubts about Putnam's program, namely the difficulties with modal accounts of the application of mathematics.

D

I would like to conclude this discussion of Putnam by saying that Putnam's views are very similar to mine in their main motivation, which is to solve the central epistemological problem that besets mathematics by offering an anti-platonist view of mathematics. Putnam and I agree on a main criterion of adequacy in an anti-platonist view: it must illuminate not only mathematics itself but also the application of mathematics, without invoking mathematical entities. And his anti-platonist view, like mine, can (though needn't necessarily) be described as saying that mathematics needn't be true to be good. I am skeptical of the details of Putnam's views, largely because I do not see how he is going to put modality to use in explaining the applicability of mathematics (as discussed in sections 6, 7 and 8A of this paper); the doubts just discussed in 8C are secondary. But I would be happy to discover that the various doubts I have made about his enterprise are misplaced: Putnam and I are trying to get to essentially the same place, even if we disagree about the best route for getting there.

Bibliography

Benacerraf, Paul 1965: 'What numbers could not be'. *Philosophical Review*, 74, 47–73.

—— 1973: 'Mathematical truth'. *Journal of Philosophy*, 19, 661–79.

Boolos, George 1984: 'To be is to be the value of a variable (or some values of some variables)'. *Journal of Philosophy*, 81, 439–49.

Burge, Tyler 1974: 'Truth and singular terms'. *Nous*, 8, 309–25.

Burgess, John 1984: 'Synthetic mechanics'. *Journal of Philosophical Logic*, 13, 379–95.

Carnap, Rudolph 1956: *Meaning and Necessity*. Chicago: University of Chicago Press.

Cartwright, Nancy 1983: *How the Laws of Physics Lie*. Oxford: Clarendon Press.

Chihara, Charles 1973: *Ontology and the Vicious Circle Principle*. Ithaca: Cornell University Press.

—— 1984: 'A simple type theory without platonic domains'. *Journal for the Philosophy of Logic*, 13, 249–83.

Cochiarella, Nino 1975: 'On the primary and secondary semantics of logical necessity'. *Journal of Philosophical Logic*, 4, 13–27.

Crossley, John and Lloyd Humberstone 1977: 'The logic of "actually"'. *Reports on Mathematical Logic*, 8, 11–29.

Davidson, Donald 1965: 'Theories of meaning and learnable languages'. In *Logic, Methodology, and Philosophy of Science*, ed. Yehoshua Bar-Hillel. Amsterdam: North-Holland, 383–94.

Detlefsen, Michael 1986: *Hilbert's Program: an essay on mathematical instrumentalism*. Dordrecht: Reidel.

Devitt, Michael 1983: 'Realism and the renegade Putnam'. *Nous*, 17, 291–301.

Dummett, Michael 1976: 'What is a theory of meaning? (II)'. In *Truth and Meaning: essays in semantics*, ed. Gareth Evans and John McDowell. Oxford: Oxford University Press, 67–137.

—— 1978: *Truth and Other Enigmas*. Cambridge, Mass.: Harvard University Press.

Feyerabend, Paul 1975: *Against Method*. London: New Left Books.

—— 1970: *Philosophy of Logic*. Englewood Cliffs: Prentice-Hall.

Ramsey, Frank 1925: 'The foundations of mathematics'. *Proceedings of the London Mathematical Society*, ser. 2, vol. 25, part V: 338–84.

Reinhardt, William 1985: 'Absolute versions of incompleteness theorems'. *Nous*, 19, 317–46.

Resnik, Michael 1980: *Frege and the Philosophy of Mathematics*. Ithaca: Cornell University Press.

—— 1983: 'Review of Field (1980)'. *Nous*, 27, 514–19.

—— 1985a: 'How nominalist is Hartry Field's nominalism?'. *Philosophical Studies*, 47, 163–81.

—— 1985b: 'Ontology and logic: remarks on Hartry Field's anti-platonist philosophy of mathematics'. *History and Philosophy of Logic*, 6, 191–209.

Scott, Dana 1967: 'Existence and description in formal logic'. In *Bertrand Russell: philosopher of the century*, ed. Ralph Schoenman. Boston: Little Brown, 181–200.

—— 1971: 'On engendering an illusion of understanding'. *Journal of Philosophy*, 68, 787–807.

Shapiro, Stewart 1983: 'Conservativeness and incompleteness'. *Journal of Philosophy*, 80, 521–31.

—— 1984: 'Review of Field (1980)'. *Philosophia*, 14, 437–44.

Sklar, Lawrence 1974: *Space, Time and Spacetime*. Berkeley: University of California Press.

Stalnaker, Robert 1976: 'Possible worlds'. *Nous*, 10, 65–75.

—— 1987: 'Critical notice: *On the Plurality of Worlds* by David Lewis'. *Mind*, 97, 117–28.

Steiner, Mark 1975: *Mathematical Knowledge*. Ithaca: Cornell University Press.

Tait, William 1986: 'Truth and proof: the platonism of mathematics'. *Synthese*, 69, 341–70.

Tarski, Alfred 1956: 'The concept of logical consequence'. In *Logic, Semantics and Metamathematics: papers from 1923 to 1938*, trans. J. H. Woodger, Oxford: Clarendon Press, 409–20.

Teller, Paul (forthcoming): 'Space-time as a physical quantity'. In the *Kelvin Memorial Volume*, ed. Robert Kargon and Peter Achinstein. Cambridge, Mass.: MIT/Bradford.

van Fraassen, Bas 1975: 'Critical notice: Hilary Putnam's *Philosophy of Logic*'. *Canadian Journal of Philosophy*, 4, 731–43.

—— 1980: *The Scientific Image*. Oxford: Clarendon Press.

van Inwagen, Peter 1981: 'Why I don't understand substitutional quantification'. *Philosophical Studies*, 38, 281–6.

Wagner, Stephen 1982: 'Arithmetical Fiction'. *Pacific Philosophical Quarterly*, 63, 255–69.

White, Nicholas 1974: 'What numbers are'. *Synthese*, 27, 111–24.

Wright, Crispin 1983: *Frege's Conception of Numbers as Objects*. Aberdeen: Aberdeen University Press.

—— forthcoming: 'Why numbers can believably be'. Manuscript. To appear in *Revue Internationale de Philosophie*.

Index